# NATURAL FOOD PRODUCTS AND WASTE RECOVERY

## Healthy Foods, Nutrition Design, and Extraction of Valuable Compounds

# NATURAL FOOD PRODUCTS AND WASTE RECOVERY

## Healthy Foods, Nutrition Design, and Extraction of Valuable Compounds

*Edited by*

**Elizabeth Carvajal-Millan, PhD**
**Abu Zahrim Yaser, PhD**
**A. K. Haghi, PhD**

APPLE
ACADEMIC
PRESS

First edition published 2022

**Apple Academic Press Inc.**
1265 Goldenrod Circle, NE,
Palm Bay, FL 32905 USA

4164 Lakeshore Road, Burlington,
ON, L7L 1A4 Canada

**CRC Press**
6000 Broken Sound Parkway NW,
Suite 300, Boca Raton, FL 33487-2742 USA

2 Park Square, Milton Park,
Abingdon, Oxon, OX14 4RN UK

© 2022 Apple Academic Press, Inc.

*Apple Academic Press exclusively co-publishes with CRC Press, an imprint of Taylor & Francis Group, LLC*

**Library and Archives Canada Cataloguing in Publication**

Title: Natural food products and waste recovery : healthy foods, nutrition design, and extraction of valuable compounds / edited by Elizabeth Carvajal-Millan, PhD, Abu Zahrim Yaser, PhD, A.K. Haghi, PhD.

Names: Carvajal-Millan, Elizabeth, editor. | Yaser, Abu Zahrim, editor. | Haghi, A. K., editor.

Description: First edition. | Includes bibliographical references and index.

Identifiers: Canadiana (print) 20210123834 | Canadiana (ebook) 20210123923 | ISBN 9781771889810 (hardcover) | ISBN 9781774638293 (softcover) | ISBN 9781003144748 (ebook)

Subjects: LCSH: Functional foods. | LCSH: Natural foods. | LCSH: Nutrition. | LCSH: Food waste—Prevention. | LCSH: Bioactive compounds.

Classification: LCC QP144.F85 N38 2021 | DDC 613.2—dc23

**Library of Congress Cataloging-in-Publication Data**

Names: Carvajal-Millan, Elizabeth, editor. | Yaser, Abu Zahrim, editor. | Haghi, A. K., editor.

Title: Natural food products and waste recovery : healthy foods, nutrition design, and extraction of valuable compounds / edited by Elizabeth Carvajal-Millan, PhD, Abu Zahrim Yaser, PhD, A. K. Haghi, PhD.

Description: Palm Bay, FL, USA : Apple Academic Press, 2021. | Includes bibliographical references and index. | Summary: "Natural Food Products and Waste Recovery: Healthy Foods, Nutrition Design, and Extraction of Valuable Compounds addresses important issues in the design of functional foods and nutraceuticals, extraction of essential compounds, and food waste management. Topics in the nutrition section cover a diverse range of topics, including uses and regulations of functional foods and ingredients, supplements, nutraceuticals, and superfoods; informatics and methods in nutrition design and development; and molecular modeling techniques in food and nutrition development. The volume goes on to address properties, microstructural characteristics, and extraction techniques of bioactive compounds. Chapters also cover the use of artificial intelligence and machine learning in food waste management, mitigation, and reuse strategies for food waste. This research-based volume is a valuable reference for professionals involved in product development and researchers focusing on food products. It will be of great interest to postgraduate students and researchers in environmental policy and waste management, as well as policymakers and practitioners in consumer issues and business"-- Provided by publisher.

Identifiers: LCCN 2021003388 (print) | LCCN 2021003389 (ebook) | ISBN 9781771889810 (hardcover) | ISBN 9781774638293 (paperback) | ISBN 9781003144748 (ebook)

Subjects: LCSH: Food industry and trade--Waste minimization. | Food industry and trade--By-products. | Food waste--Recycling. | Agricultural wastes. | Biological products.

Classification: LCC TP373.8 .N38 2021 (print) | LCC TP373.8 (ebook) | DDC 664/.08--dc23

LC record available at https://lccn.loc.gov/2021003388

LC ebook record available at https://lccn.loc.gov/2021003389

ISBN: 978-1-77188-981-0 (hbk)
ISBN: 978-1-77463-829-3 (pbk)
ISBN: 978-1-00314-474-8 (ebk)

# About the Editors

**Elizabeth Carvajal-Millan, PhD**
*Research Scientist, Research Center for Food and Development (CIAD), Hermosillo, Mexico*

Elizabeth Carvajal-Millan, PhD, is a Research Scientist at the Research Center for Food and Development (CIAD) in Hermosillo, Mexico, since 2005. She obtained her PhD in France at Ecole Nationale Supérieure Agronomique à Montpellier (ENSAM), her MSc degree at CIAD, and her undergraduate degree at the University of Sonora in Mexico. Her research interests are focused on biopolymers, mainly in the extraction and characterization of high-value-added polysaccharides from co-products recovered from the food industry and agriculture, especially ferulated arabinoxylans (AXs). In particular, Dr. Carvajal-Millan studies covalent arabinoxylans gels as functional systems for the food and pharmaceutical industries. Globally, Dr. Carvajal-Millan is a pioneer in *in vitro* and *in vivo* studies on covalent arabinoxylans gels as carriers for oral insulin focused on the treatment of diabetes type 1. She has published over 60 refereed papers, more than 20 chapters in books, and over 80 conference presentations. She holds one patent registered, with two more submitted.

**Abu Zahrim Yaser, PhD**
*Senior Lecturer, Chemical Engineering Program, University Malaysia Sabah (UMS), Malaysia*

Abu Zahrim Yaser, PhD, is Associate Professor at the Chemical Engineering Programme, Universiti Malaysia Sabah (UMS). He obtained his PhD from Swansea University (UK), MSc from Universiti Kebangsaan Malaysia (Malaysia), and BEng from the University of Malaya (Malaysia). Dr. Zahrim's research mainly focuses on waste processing technology, and he has been leading several projects since 2006. To date, Dr. Zahrim has published six books, over 19 book chapters, over35 journals and over 50 other publications. He was the Guest Editor for a special issue of Environmental Science and Pollution Research (Springer). He has won several medals in innovation competitions, including a gold medal at iENA 2018, Germany.

His innovation entitled "UMS Organic Compost" helps Sabah state to win the "Best State Innovation Award" at IGEM 2019. His invention, the Auto Turning Active Zone Yield Composter (Auto AZY), has been adopted by the Tongod District Council, Sabah, for managing their organic waste. Dr. Zahrim has reviewed more than 80 manuscripts (based on Publons) and the Elsevier (UK) has recognized him as among the Outstanding Reviewers for Journal of Environmental Chemical Engineering (2018). He was the co-chair for the International Symposium on Carbon and Functional Materials for Energy and Environment (C-MEE 2020). Dr. Zahrim was a visiting scientist at the University of Hull. Dr. Zahrim is a member of Institutions of Chemical Engineers (United Kingdom), Board of Engineers, Malaysia and MyBIOGAS.

### A. K. Haghi, PhD

*Professor Emeritus of Engineering Sciences, Former Editor-in-Chief, International Journal of Chemoinformatics and Chemical Engineering and Polymers Research Journal; Member, Canadian Research and Development Center of Sciences and Culture*

A. K. Haghi, PhD, is the author and editor of over 200 books, as well as over 1000 published papers in various journals and conference proceedings. Dr. Haghi has received several grants, consulted for a number of major corporations, and is a frequent speaker to national and international audiences. Since 1983, he served as professor at several universities. He is the former Editor-in-Chief of the *International Journal of Chemoinformatics and Chemical Engineering* and *Polymers Research Journal* and is on the editorial boards of many international journals. He is also a member of the Canadian Research and Development Center of Sciences and Cultures. He holds a BSc in urban and environmental engineering from the University of North Carolina (USA), an MSc in mechanical engineering from North Carolina A&T State University (USA), a DEA in applied mechanics, acoustics and materials from the Université de Technologie de Compiègne (France), and a PhD in engineering sciences from Université de Franche-Comté (France).

# Contents

# Contributors

**Cristóbal Noé Aguilar**
Bioprocesses and Bioproducts Research Group, Food Research Department, School of Chemistry, Autonomous University of Coahuila, Saltillo – 25280, México, E-mail: cristobal.aguilar@uadec.edu.mx

**Agustín Rascón-Chu,**
Biotechnology-CTAOV. Research Center for Food and Development (CIAD, A.C.), Carretera Gustavo Enrique Astiazarán Rosas No. 46, Hermosillo – 83304, Sonora, Mexico

**Eknath D. Ahire**
Divine College of Pharmacy, Nampur Road, Satana, Maharashtra, India

**S. Y. Ang**
National Hydraulic Research Institute of Malaysia NAHRIM, 43300 – Seri Kembangan, Selangor, Malaysia

**Roberto Arredondo-Valdés**
Nanobioscience Group, Chemistry School, Autonomous University of Coahuila, Saltillo, Mexico

**J. A. Ascacio-Valdés**
Bioprocesses and Bioproducts Group, Food Research Department, School of Chemistry, Autonomous University of Coahuila, Saltillo – 25280, Coahuila, México, E-mail: alberto_ascaciovaldes@uadec.edu.mx

**Mairuz Asmarafariza Azlan**
Zero Waste Campaign, University Malaya, Kuala Lumpur – 50603, Malaysia, E-mail: mairuzasmara@gmail.com

**Sitty Nur Syafa Bakri**
Preparatory Center for Science and Technology, University Malaysia Sabah, Jalan UMS, Kota Kinabalu – 88400, Sabah, Malaysian, E-mail: syafa@ums.edu.my

**Razsera Hassan Basri**
Development Division, IIUM Gombak, Malaysia

**Ena D. Bolaina-Lorenzo**
Laboratory of Applied Glycobiotechnology, Food Research Department, School of Chemistry, Autonomous University of Coahuila, Saltillo – 25250, Coahuila, Mexico

**Francisco Brown-Bojorquez**
University of Sonora, Rosales y Blvd, Luis D. Colosio, Hermosillo, Sonora – 83000, Mexico

**Alma Campa-Mada**
Biopolymers-CTAOA, Research Center for Food and Development (CIAD, A.C.), Carretera Gustavo Enrique Astiazarán Rosas No. 46, Hermosillo – 83304, Sonora, Mexico

**Carlos N. Cano-Gonzalez**
School of Chemistry, Autonomous University of Coahuila, Saltillo – 25280, Coahuila, Mexico

**Elizabeth Carvajal-Millan**
Biopolymers-CTAOA, Research Center for Food and Development (CIAD, A.C),
Carretera Gustavo Enrique Astizaran Rosas No. 46, Hermosillo, Sonora – 83304, Mexico,
Tel.: +52-662-289-2400, Fax: +52-662-280-0421, E-mail: ecarvajal@ciad.mx

**Maria I. Castillo-Sanchez**
Laboratory of Applied Glycobiotechnology, Food Research Department,
Department of Chemical Engineering, School of Chemistry, Autonomous University of Coahuila,
Saltillo – 25250, Coahuila, Mexico

**Mónica Lizeth Chávez-González**
Nanobioscience Group, Chemistry School, Autonomous University of Coahuila, Saltillo,
Mexico; Bioprocesses and Bioproducts Research Group, Food Research Department,
Autonomous University of Coahuila, Saltillo – 25280, Mexico

**Agustín Rascón Chu**
Research Center for Food and Development (CIAD, A.C), Carretera Gustavo Enrique Astizaran Rosas
No. 46, Hermosillo, Sonora – 83304, Mexico

**Gabriela P. Cid-Ibarra**
School of Chemistry, Autonomous University of Coahuila, Saltillo – 25280, Coahuila, Mexico

**Juan Carlos Contreras-Esquivel**
Laboratory of Applied Glycobiotechnology, Food Research Department, School of Chemistry,
Autonomous University of Coahuila, Saltillo – 25250, Coahuila, México,
E-mail: carlos.contreras@uadec.edu.mx

**Pranjal Pratim Das**
Department of Chemical Engineering, IIT Guwahati, Assam, India

**Elnetthra Folly Eldy**
Preparatory Center for Science and Technology, University Malaysia Sabah, Jalan UMS,
Kota Kinabalu – 88400, Sabah, Malaysian

**Jorge Alberto Márquez Escalante**
Research Center for Food and Development (CIAD, A.C), Carretera Gustavo Enrique Astizaran Rosas
No. 46, Hermosillo, Sonora – 83304, Mexico

**Sandra Cecilia Esparza-González**
Odontology Faculty, Autonomous University of Coahuila, Saltillo, Mexico

**Luis E. Estrada-Gil**
Bioprocesses and Bioproducts Group, Food Research Department, School of Chemistry,
Autonomous University of Coahuila, Saltillo – 25280, Coahuila, México

**Eva L. Fernández-Rodríguez**
School of Chemistry, Autonomous University of Coahuila, Saltillo – 25280, Coahuila, Mexico

**José Miguel Fierro-Islas**
Research Center for Food and Development (CIAD, A.C), Carretera Gustavo Enrique Astizaran Rosas
No. 46, Hermosillo, Sonora – 83304, Mexico

**Adriana C. Flores-Gallegos**
Bioprocesses and Bioproducts Group, Food Research Department, School of Chemistry,
Autonomous University of Coahuila, Saltillo – 25280, Coahuila, México

**Ramesh Gadekar**
Sandip Institute of Pharmaceutical Sciences, Mahiravani, Nasik, Maharashtra, India

**Jesús Alberto García García**
School of Education, Sciences and Humanities, Autonomous University of Coahuila, Saltillo, México, E-mail: jegarciag@uadec.edu.mx

**J. Daniel García-García**
Nanobioscience Group, Chemistry School, Autonomous University of Coahuila, Saltillo, Mexico

**Melany García-Moreno**
Nanobioscience Group, Chemistry School, Autonomous University of Coahuila, Saltillo, Mexico

**Mayela Govea-Salas**
Laboratory of Nanobioscience, School of Chemistry, Autonomous University of Coahuila, Saltillo – 25280, Coahuila, México; Bioprocesses and Bioproducts Research Group, Food Research Department, Autonomous University of Coahuila, Saltillo – 25280, Mexico

**Ng Chee Guan**
Institute of Ocean and Earth Science, University Malaya, Kuala Lumpur – 50603, Malaysia, E-mail: cheeguan.ng@um.edu.my

**Viviana C. Guillermo-Balderas**
Laboratory of Applied Glycobiotechnology, Food Research Department, School of Chemistry, Autonomous University of Coahuila, Saltillo – 25250, Coahuila, Mexico

**Manuj Kumar Hazarika**
Department of Food Engineering and Technology, Tezpur University, Assam, India

**Mohd. Ramzi Mohd. Hussain**
Associate Professor, Kulliyyah of Architecture and Environmental Design, IIUM Gombak, Malaysia

**M. R. M. Huzaifah**
Department of Crop Science, Faculty of Agriculture, Universiti Putra Malaysia (UPM), 43400 – UPM Serdang, Selangor, Malaysia, E-mail: mdhuzaifahroslim@gmail.com

**Anna Ilyina**
Nanobioscience Group, Chemistry School, Autonomous University of Coahuila, Saltillo, Mexico, Tel.: +52 (844) 139-6175, E-mail: annailina@uadec.edu.mx

**Iziana Hani Ismail**
Preparatory Center for Science and Technology, University Malaysia Sabah, Jalan UMS, Kota Kinabalu – 88400, Sabah, Malaysian

**S. Ismail**
School of Ocean Engineering, Universiti Malaysia Terengganu, 21030 – Kuala Nerus, Terengganu, Malaysia

**Anil Jadhav**
Sandip Institute of Pharmaceutical Sciences, Mahiravani, Nasik, Maharashtra, India

**Jaime Lizardi-Mendoza**
Biopolymers-CTAOA, Research Center for Food and Development (CIAD, A.C.), Carretera Gustavo Enrique Astiazarán Rosas No. 46, Hermosillo – 83304, Sonora, Mexico

**A. Jamali**
Department of Mechanical and Manufacturing Engineering, Faculty of Engineering, Jalan Datuk Mohd. Musa – 94300, University Malaysia Sarawak, Malaysia, E-mail: jannisa@unimas.my

**Z. Johar**
National Hydraulic Research Institute of Malaysia NAHRIM, 43300 – Seri Kembangan, Selangor, Malaysia

**Judith Tanori-Cordova**
Department of Polymers and Materials Research, University of Sonora, Hermosillo – 83000, Sonora, Mexico

**Dato' Wan Mohd. Hilmi Wan Kamal**
Development Division, IIUM Gombak, Malaysia

**Liliana Londoño-Hernandez**
Applied Microbiology and Biotechnology Research Group, Department of Biology, Universidad del Valle, Cali, Colombia

**Rodrigo Macias-Garbet**
Laboratory of Applied Glycobiotechnology, Food Research Department, School of Chemistry, Autonomous University of Coahuila, Saltillo – 25250, Coahuila, Mexico

**Ana L. Martínez-López**
NANO-VAC Research Group, Department of Chemistry and Pharmaceutical Technology, University of Navarra, Pamplona – 31008, Spain

**José Luis Martínez-Hernández**
Bioprocesses and Bioproducts Research Group, Food Research Department, Autonomous University of Coahuila, Saltillo – 25280, Mexico; Nanobioscience Group, Chemistry School, Autonomous University of Coahuila, Saltillo, Mexico

**Marcel Martínez-Porchas**
Research Center for Food and Development (CIAD, A.C), Carretera Gustavo Enrique Astizaran Rosas No. 46, Hermosillo, Sonora – 83304, Mexico

**Karla Martínez-Robinson**
Research Center for Food and Development (CIAD, A.C), Carretera Gustavo Enrique Astizaran Rosas No. 46, Hermosillo, Sonora – 83304, Mexico

**Marco A. Mata-Gómez**
School of Engineering and Science, Tecnológico de Monterrey, Puebla – 5718, 72453, México

**A. N. Mazlan**
University Malaysia Terengganu, Kuala Nerus – 21030, Terengganu, Malaysia

**I. M. Mehedi**
Center of Excellence in Intelligent Engineering Systems (CEIES), King Abdulaziz University, Jeddah, Saudi Arabia

**Emilio Mendez-Merino**
Sigma Alimentos Co., Torre Sigma, San Pedro Garza Garcia – 66269, Nuevo Leon, Mexico

**Z. F. A. Misman**
Department of Mechanical and Manufacturing Engineering, Faculty of Engineering, Jalan Datuk Mohd. Musa – 94300, University Malaysia Sarawak, Malaysia

**Julio C. Montañez**
Department of Chemical Engineering, School of Chemistry, Autonomous University of Coahuila, Saltillo City – 25250, Coahuila, Mexico

**Paulina V. Parga-Hernandez**
Laboratory of Applied Glycobiotechnology, Food Research Department, School of Chemistry, Autonomous University of Coahuila, Saltillo – 25250, Coahuila, Mexico

**Erick M. Peña-Lucio**
Bioprocesses and Bioproducts Research Group, Food Research Department, Autonomous University of Coahuila, Saltillo – 25280, Mexico

**Brenda C. Ramirez-Gonzalez**
Laboratory of Applied Glycobiotechnology, Food Research Department, School of Chemistry, Autonomous University of Coahuila, Saltillo – 25250, Coahuila, Mexico

**Ana Mayela Ramos-De-La-Peña**
Tecnologico de Monterrey, School of Engineering and Science. Av. Eugenio Garza Sada – 2501 Sur, Monterrey, Nuevo Leon – 64849, Mexico, E-mail: ramos.amay@tec.mx

**Rodolfo Ramos-González**
Nanobioscience Group, Chemistry School, Autonomous University of Coahuila, Saltillo, Mexico

**Merab Magaly Rios-Licea**
Sigma Alimentos Co., Torre Sigma, San Pedro Garza Garcia – 66269, Nuevo Leon, Mexico

**María G. Rodríguez-Delgado**
School of Chemistry, Autonomous University of Coahuila, Saltillo – 25280, Coahuila, Mexico

**Raúl Rodríguez-Herrera**
Laboratory of Molecular Biology, Food Research Department, School of Chemistry, Autonomous University of Coahuila, Saltillo – 25280, Coahuila, México

**Rafael Canett Romero**
University of Sonora, Rosales y Blvd, Luis D. Colosio, Hermosillo, Sonora – 83000, Mexico

**Hector Ruiz-Leza**
Bioprocesses and Bioproducts Research Group, Food Research Department, Autonomous University of Coahuila, Saltillo – 25280, Mexico

**Abdulhameed Sabu**
Department of Biotechnology and Microbiology, Kannur University, Thalassery Campus, Kannur – 670661, India

**Mohd. Armi Abu Samah**
Assistant Professor, Kulliyyah of Science, IIUM Kuantan Campus, Malaysia

**Adriana J. Sañudo-Barajas**
Research Center for Food and Development, AC, Culiacan – 80110, Sinaloa, Mexico

**Elda Patricia Segura-Ceniceros**
Nanobioscience Group, Chemistry School, Autonomous University of Coahuila, Saltillo, Mexico

**N. A. Sheikh**
Department of Mechanical Engineering, Faculty of Engineering, International Islamic University, Islamabad, Pakistan

**Nikita Shevkar**
Sandip Institute of Pharmaceutical Sciences, Mahiravani, Nasik, Maharashtra, India

**Vijayraj N. Sonawane**
Divine College of Pharmacy, Nampur Road, Satana, Maharashtra, India

**Khemchand R. Surana**
Divine College of Pharmacy, Nampur Road, Satana, Maharashtra, India

**G. S. Talele**
Matoshri College of Pharmacy, Eklhre Nasik, Maharashtra, India

**Swati G. Talele**
Sandip Institute of Pharmaceutical Sciences, Mahiravani, Nasik, Maharashtra, India,
E-mail: swatitalele77@gmail.com

**Preetisagar Talukdar**
Department of Chemical Engineering, IIT Guwahati, Assam, India,
E-mail: preetisagar1891@gmail.com

**K. F. Tamrin**
Department of Mechanical and Manufacturing Engineering, Faculty of Engineering,
University Malaysia Sarawak (UNIMAS), Kota Samarahan – 94300, Sarawak, Malaysia,
Tel.: +601115653090, Fax: +6082583410, E-mail: tkfikri@unimas.my

**N. A. Umor**
Faculty of Applied Sciences, School of Biology, Universiti Teknologi MARA (UiTM) Negeri Sembilan,
Kampus Kuala Pilah, Pekan Parit Tinggi, Kuala Pilah – 72000, Negeri Sembilan, Malaysia,
Phone: +601110910939, Fax: +603-8940-8319, E-mail: noorazrimi@gmail.com

**Oscar F. Vazquez-Vuelvas**
Faculty of Chemical Sciences, Universidad de Colima, Coquimatlan – 28400, Colima, Mexico

**Rosabel Velez-De-La-Rocha**
Research Center for Food and Development, AC, Culiacan – 80110, Sinaloa, Mexico

**F. Yasmin**
Department of Mechanical and Manufacturing Engineering, Faculty of Engineering,
University Malaysia Sarawak (UNIMAS), Kota Samarahan – 94300, Sarawak, Malaysia

**Yubia B. De Anda-Flores**
Biopolymers-CTAOA, Research Center for Food and Development (CIAD, A.C),
Carretera Gustavo Enrique Astiazarán Rosas No. 46, Hermosillo – 83304, Sonora, Mexico

**Kamaruzzaman Yunus**
Professor, Kulliyyah of Science, IIUM Kuantan Campus, Malaysia

**Sumiani Yusoff**
Institute of Ocean and Earth Science, University Malaya, Kuala Lumpur – 50603, Malaysia,
E-mail: sumiani@um.edu.my

**M. S. M. Zahari**
University Malaysia Terengganu, Kuala Nerus – 21030, Terengganu, Malaysia

**Nur Shakirah Binti Kamarul Zaman**
Institute of Advanced Studies, University Malaya, Kuala Lumpur – 50603, Malaysia,
E-mail: shakirahkz90@gmail.com

**Alejandro Zugasti-Cruz**
Laboratory of Immunology, School of Chemistry, Autonomous University of Coahuila,
Saltillo – 25280, Coahuila, México

**A. Zulqarnain**
University Malaysia Terengganu, Kuala Nerus – 21030, Terengganu, Malaysia

# Abbreviations

| | |
|---|---|
| A/X | arabinose/xylose |
| ABS | acrylonitrile butadiene styrene |
| AD | anaerobic digestion |
| AI | artificial intelligence |
| AIDS | acquired immune deficiency syndrome |
| AlTiN | aluminum titanium nitride |
| ANN | artificial neural network |
| AXs | arabinoxylans |
| BSF | black soldier fly |
| BSFL | black soldier fly larvae |
| BSG | Brewer's spent grain |
| CBIR | content-based image retrieval |
| CCD | central composite design |
| CEL | $N\varepsilon$-(carboxyethyl)lysine |
| CML | $N\varepsilon$-(carboxymethyl)lysine |
| CMP | cyclic adenosine monophosphate |
| CND | construction and demolition |
| CNN | convolutional neural network |
| ConA | concanavalin A |
| CP | composting |
| CT | complementary troubleshooting |
| DBSB | Daya Bersih Sdn. Bhd. |
| DDGS | distillers dried grains with solubles |
| DFT | density functional theory |
| DHA | docosahexaenoic acid |
| di-FA | dehydrodimers of ferulic acid |
| DNA | deoxyribonucleic acid |
| ECD | efficiency of conversion of digested |
| ECG | epicatechin gallate |
| EFA | essential fatty acids |
| EGCG | epigallocatechin gallate |
| ELM | extreme learning machine |
| ELVs | end of life vehicles |
| EM | European mistletoe |

| | |
|---|---|
| EMC | equilibrium moisture content |
| EPA | eicosapentaenoic acid |
| EU | European Union |
| EW | e-waste |
| FA | ferulic acid |
| FAO | Food and Agricultural Organization |
| FDA | Food and Drug Administration |
| FFNN | feedforward neural network |
| FOSHU | foods for specified health use |
| FSU | food and services unit |
| FT-IR | Fourier transform infrared |
| FW | food waste |
| GIT | gastrointestinal tract |
| GW | green waste |
| HB | hardness Brinell |
| HDL | high-density lipoprotein |
| HIV | human immunodeficiency virus |
| HMG-CoA | 3-hydroxy-3-methylglutaryl coenzyme A |
| HMM | hidden Markov model |
| HOMO | highest energy occupied molecular orbital |
| HPLC | high performance liquid chromatography |
| IGES | global environmental strategies |
| IGF | insulin-like growth factor |
| IIUM | Islamic International University Malaysia |
| ISWM | integrated solid waste management |
| KM | Korean mistletoe |
| KML | Korean mistletoe lectin |
| KNN | k-nearest neighbor |
| KRA | key result area |
| LDL | low-density lipoprotein |
| LLC | lifeline clothing |
| LUMO | lowest energy unoccupied molecular orbital |
| LWIR | long-wave infrared |
| MFs | metallic fractions |
| MIT | Massachusets International Technology |
| MLP | multilayer perceptron |
| MLs | mistletoe lectins |
| MMW | millimeter-wave |
| MR | moisture ratio |

| | |
|---|---|
| MSM | *Minggu Suai Mesra* |
| MSW | municipal solid waste |
| Mw | molecular weight |
| NCE | new chemical entity |
| NIR | near-infrared |
| NMFs | non-metallic fractions |
| NSWMD | National Solid Waste Management Department |
| PAs | proanthocyanidins |
| PCA | principal component analysis |
| PCA-SVM | principle component analysis-support vector machine |
| PDB | protein data bank |
| PEF | pulsed electric fields |
| PET | polyethylene-terephthalate |
| PP | polypropylene |
| PPST | Preparatory Center for Science and Technology |
| PpyLL | *Phthirusa pyrifolia* leaf lectin |
| QSAR | quantitative structure-activity relationship |
| RB | recyclables |
| R-CNN | region-convolutional neural network |
| RF | random forest |
| ROSDAL | representation of organic structures description arranged linearly |
| RPN | region proposal generation |
| RWD | refractance window drying |
| RW™ | refractance window |
| SCF | self-consistent field process |
| SDGs | sustainable development goals |
| SEM | scanning electron microscopy |
| SLN | SYBYL line notation |
| SMILES | simplified molecular input line entry specification |
| SSD | spectral-spatial density |
| SSE | sum squared error |
| STn antigen | sialyl-Tn antigen |
| SVM | support vector machine |
| SWM | solid waste management |
| TEOS | tetraethyl orthosilicate |
| TLRs | toll-like receptors |
| UM ZWC | University Malaya Zero Waste Campaign |
| UM | University Malaya |

| | |
|---|---|
| UMIN | University Hospital Medical Information Network |
| UMS | University Malaysia Sabah |
| UMT | University Malaysia Terengganu |
| URS | unsupervised robotic sorting |
| UT | used textile |
| VAAs | viscum album agglutinins |
| VFAs | volatile fatty acids |
| WEAXs | water-extractable arabinoxylans |
| WEEE | waste of electric and electronic equipment |
| WHO | World Health Organization |
| WLN | Wiswesser line notation |
| WUAXs | water unextractable arabinoxylans |
| WW | wood waste |
| ZRR | ZenRobotics recycler |

# Preface

As time goes on and technology keeps on moving forward, innovation in all the fields of science keeps ongoing, and this is no exception for food science and technology, where new products are developed to satisfy new demands in our population. This generation of consumers looks to nourish with the natural nutrients that our foods give us.

Food is a product for human ingestion that provides nutritional support for healthy growth. When food is disposed or not consumed, it becomes a waste. It is a correlation between knowledge, skills, and behavior to be able to plan, manage, select, and prepare food as well as eating food.

Meanwhile, focusing on the crucial sustainability challenge of reducing food loss at the level of consumer society, this volume provides an in-depth, research-oriented overview of this multifaceted problem. This book also considers the myriad environmental, economic, social, and ethical factors associated with an enormous amount of food waste (FW), which also ends up wasting water, air, electricity, and fuel, which are necessary for food processing.

Novel health food products have been widely studied as they provide the population with a lot of health benefits within their grasp as they can be purchased in many stores and supermarkets; the only thing that is necessary for people to know the benefits of these products is for them to get the information they need.

This valuable research-based book is divided into three sections: (i) Section 1: Healthy foods and nutrition; (ii) Section 2: Extraction of essential compounds; and (iii) Section 3: Food waste management.

This volume is a valuable reference for professionals involved in product development and researchers focusing on food products and will be of great interest to postgraduate students and researchers in environmental policy and waste management, as well as policymakers and practitioners in consumer issues and business.

# PART 1
# Healthy Food and Nutrition

# CHAPTER 1

# Bovine Colostrum: Food Supplement

NIKITA SHEVKAR, RAMESH GADEKAR, SWATI TALELE, and
ANIL JADHAV

*Sandip Institute of Pharmaceutical Sciences, Mahiravani, Nasik,
Maharashtra, India, E-mail: swatitalele77@gmail.com (S. Talele)*

## ABSTRACT

Colostrum is a sticky white or yellow nutrient-rich liquid secreted by female mammals immediately afterward giving birth. Bovine colostrum is load with massive nutrients like vitamins, minerals, fats, carbohydrates, disease-fighting proteins, growth hormones, and digestive enzymes. It also possesses functional food properties. Nowadays, society is more interested in bovine colostrum supplements because they promote immunity, exchange blows infection, and get well gut health. To get ready, these supplements colostrum from a cow is pasteurized and then dries into pills or into a powder that can be mixed with liquid. Bovine colostrum possesses outstanding properties, including various antibodies such as IgA, IgG, and IgM, which are used to strengthen our immune response to fight against microorganisms including bacteria and viruses. Colostrum restrains protein-based hormones like IGF-1 and IGF-2, which stimulate the growth. This growth factor shows enormous significant muscle and cartilage repair characteristics. They encourage wound healing with real consequences for suffering and operating patients. Lactoferrin, which is a constituent of colostrum is a protein involved in our body's immune reaction to infections, containing folks triggered by bacteria and viruses. Lactoferrin also possess antiviral and antibacterial activity. Colostrum acts as a powerful inhibitor in eye infection against bacteria, i.e., *Chlamydia tracomatis*. Even colostrum is topically applied, it effectively acts against dryness of the eye and eye injury. Presently scientists are a move towards new innovation to clone and reproduce essential main factor occurs in colostrum. Researchers utilized *in vivo* and an *in vitro* model to reduce gastrointestinal damage that occurs by consuming a non-steroidal anti-inflammatory agent.

Thus this review probably focuses on the efficient utilization of colostrum in day-to-day life with a couple of medicinal properties.

## 1.1   INTRODUCTION

Colostrum bovine is the first milk concealed from the mammary glands of female mammals after pasteurization or giving birth. It is also called bisnings, beestings or first milk. Bovine colostrum is a sticky white or yellowish liquid load with massive nutrients like vitamins, minerals, fats, carbohydrates, disease-fighting proteins, growth factors, and digestive enzymes.

Bovine colostrum is a very complex bioactive constituent needed to maintenance to development and repair of cells and tissue to ensure the effective and efficient metabolism of nutrients and to strengthen the immune system. Bovine colostrum is also used as a nutraceutical product. Nutraceutical product is nothing but any food or ingredients having a nutritive value that has impact on the individuals health, physical performance or state of mind and nowadays people are more attracted towards supplements due to lack of side effects and showing extra benefits [46, 47].

Researchers have developed a special type of bovine colostrum called "hyperimmune bovine colostrum". This colostrum is collected within 24 hrs of the cow giving birth. This hyperimmune BC is created by vaccinating cows against a particular disease causing microorganism. These vaccinations cause the cows to develop antibodies to struggle against specific organism, and then these antibodies pass into the colostrum [48].

Hyperimmune bovine colostrum has been second-hand in clinical trials for treating AIDS-related diarrhea. A number of athletes consume bovine colostrum by mouth to burn fat, build lean muscle, increase grit, and liveliness, and increase athletic performance [48].

**Dosage:**

> **For Adult:**

1. For infectious diarrhea-10–100 gm. of BC daily for 1–4 weeks.
2. For the flu-400 mg of freeze-dried BC daily for 8 weeks.

> **For Children (by mouth)**

1. 3 gm. of bovine colostrum once daily for 4 weeks for children having age under 1–2 years.
2. 3 gm. twice daily for 4 weeks children having age 2–6 year.

### 1.1.1   COMPONENTS OF COLOSTRUM

The components of colostrum are categorized into four categories. These components of colostrum are given away in Figure 1.1.

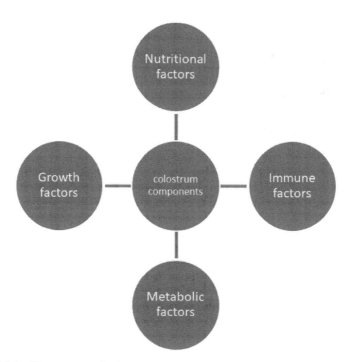

**FIGURE 1.1**   Components of colostrums.

1. **Immune Factors:** These contain components given in Table 1.1.

**TABLE 1.1**   Immune Factors [10, 14, 15]

| Components | Uses |
| --- | --- |
| **1. Immunoglobulins** | |
| i.   IgG | IgG neutralizes toxins and bacteria in the lymph and circulatory system. |
| ii.  IgM | IgM kills the bacteria. |
| iii. IgE and IgD | Highly antiviral. |
| 2. Lactoferrin | As antiviral, anti-inflammatory, and antibacterial. |
| 3. Lysosome | Improve the immune system and capable of kills bacteria and virone on contact. |
| 4. Protein-rich polypeptide | Regulate thymus gland. |

2. **Growth Factors:** These are classified as insulin-like growth factor (IGF) and growth hormone [14–17]:
   - **Insulin:** Like growth factors, IGFs are potent hormones that are found in all cells in the body. These are categorized into three types [14–17]:
     - **Transforming Growth Factor:** These brings the transformation of cells from an immature form to a mature.
     - **Epithelial Growth Factor:** It involved in the generation and maintenance of cells in epithelial layers of skin; and
     - **Platelet-Derived Growth Factor:** It helps in the generation of cells and function associated with blood clotting.
   - **Growth Hormones:** A minor quantities of growth hormone are set up in colostrum, but that is all that is essential since this hormone is very potent. It has a direct effect on every cell type and significantly influences the proliferation of new cells, mainly their rate of generation [14–17].
3. **Nutritional Components:** These components and their uses are given in Table 1.2.

**TABLE 1.2**   Nutritional Components [16–18]

| Components | Uses |
| --- | --- |
| Nutritional components vitamins (A, B12 and E) Minerals Amino Acids Essential oil | Health and growth of newborn |

4. **Metabolic Factor:** This in BC, includes the following components [10, 16–18]:
   - **Leptin:** Hormone like proteins can suppress appetite and enhance metabolic rate, which leads to a reduction in body weight.
   - **Insulin:** It is required for the effective metabolic utilization of glucose.
   - **Vitamin Binding Protein:** Act as carriers to carry B complex vitamins to the body.
   - **Cyclic Adenosine Monophosphate (CMP):** It is applicable to energy transmission in metabolism.
   - **Enzyme Inhibitors:** Inhibit the breakdown of proteins by several enzymes. This also provides protection to the growth, immune, and metabolic factor which are pass through the digestive tract.

## 1.2   METHODS

1. **Manufacturing of Colostrum Powder:** The manufacturing of colostrum powder involves three steps [45]. These three steps are shown in Figure 1.2.

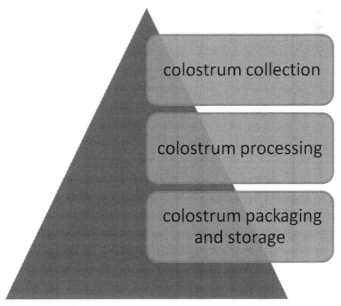

**FIGURE 1.2**   Steps of manufacturing of colostrum powder.

- **Step 1:** Collection of Colostrum: Colostrum is received from dairy farms or dairy plant that are approved to produce milk for human being consumption. Colostrum has got to be serene in the head milking within the 6 hrs after birth of the calf. Gather it into a sterilized cans and deep-rooted freezers to the minute icy the massiveness colostrum. Formerly the product is frozen, it is sent to the dispensation facility to testing for quality before accepting it for final processing [45].
- **Step 2:** Colostrum Processing: The colostrum processing involves two steps to ensure product purity, potency, safety, and stability. The process of preparation of colostrum powder is shown in Figure 1.3.

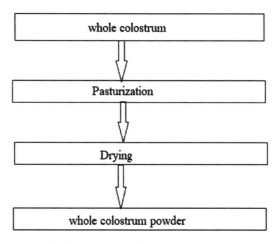

**FIGURE 1.3**  Flowchart of colostrum processing.

Processing of colostrum contains two important steps:

- ○ **Pasteurization:** Harvested colostrum is then subjected to the pasteurization. It is pasteurized as 60°C for 30 min. Pasteurization is the heat treatment process that destroys pathogenic microorganism in the colostrum. Both heating and cooling processes are strictly controls to obtain a quality product [42, 43].
- ○ **Drying:** After the completion of the pasteurization step, the colostrum is pass to the drying procedure. The colostrum is dried by two processes [42, 43] shown in Figure 1.4:

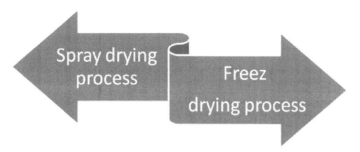

**FIGURE 1.4**   Drying process.

i. **Spray Drying Process:** It allows a liquid product to be atomized and contacted with hot gas to obtain a powder. The pasteurized colostrum is passed through a centrifugal pump on to the drying chamber. The spray at the higher of the drying chamber exchanges the colostrum liquid into a minute droplets and spray at high-level velocity inside the chamber which contains pungent air. As the droplets dry out midstream, the colostrum powder cataract at the bottom of the chamber where it is packed aseptically under precise environments. In spray drying process, dietary cherish of the product is slightly lose because it involves the use of heat [42, 43].

ii. **Freeze Drying Process:** It involves freezing of the product and vaporizing the moisture from the product to get powder. In this process, the heat is not utilized; hence the nutritive value of the product is preserved [42, 43].

• **Step 3:** Packaging and Storage of Colostrum Powder: After formation of the colostrum powder, it once again checked for purity and potency and then packed into the container. Colostrum is stored at 4° in a plastic container to maintain its protective quality. Powder forms of colostrum are packed in polyethylene-terephthalate (PET) pouches. Colostrum powder is also packaged in 200 gm. of HDPE bags and stored in an airtight tin container at 30+ –2°C. Shelf life of colostrum powder is about 6 to 24 months if stored under specified conditions [44].

## 1.3   APPLICATIONS OF BOVINE COLOSTRUM

Colostrum has many therapeutic uses. It is also used as a nutritional supplement, and subsequently, it is well abided. Colostrum is used in the following diseases:

1. **Wound Healing:** Colostrum provides excellent immune support. It contains two growth factors, i.e., alpha, and beta transforming growth factor and insulin-like growth factor 1 and 2 (IGF 1 and 2). These growth factors shows enormous significant muscle and cartilage repair characteristics and stimulate wound healing with practical consequences for trauma and surgical patients [1, 7, 8].

2. **Allergies and Autoimmune Diseases:** Allergy is also called hypersensitivity reaction or disorder of immune system. Hypersensitivity reaction arises due to incorrect immune response to hormone to the substances. Colostrum contains regulatory substance like proline-rich polypeptide which regulates the thymus gland.

   Proline-rich polypeptides also eliminate symptoms of autoimmune diseases such as rheumatoid arthritis, myasthenia gravis, sclerosis, and allergies [11, 12].

3. **Cardiovascular Disease:** Colostrum has a valuable effect in the inhibition of heart disease due to the presence of proline-rich polypeptide, the same consequence is observed in allergy and autoimmune diseases. Colostrum contains growth hormone and growth factors such as IGF can increase the blood level of HDL (high-density lipoprotein) cholesterol, while decrease LDL (low-density lipoprotein) cholesterol. Both the growth factor and growth hormone shows an important role in restoring the damage to heart muscle and support the growth of new blood vessels in the collateral coronary circulation. Bovine colostrum possesses concentration associated antioxidant activity as shown by free radical scavenging ability and to prevent the inhibition of lipid peroxidation. The colostrum also shows the cardio-protective effect [40].

4. **Diabetes:** Type-1 or insulin-dependent (Juvenile diabetes) is thought to arise through an autoimmune disorder, primarily inhabited by an intense allergic response to the glutamic acid decarboxylase protein found in cow's milk. colostrum contains bioactive factor such as immunoglobulin which control and inhibit to autoimmune disorder and decrease glucose level in blood [37].

5. **Anticancer Activity:** Bovine colostrum contains one constituent named lactalbumine, which is responsible for the physiological cell death of cancerous cells (apoptosis). colostrum also contains lactoferrin anticancer substance. This lactoferrin increase the treatment responsiveness and immune response through Th1 and Th2 activation and also increase the leukocyte and erythrocyte count. Cancer

metastasis can be inhibited by the growth and immune factor present in colostrum. NK cells found in colostrum provide resistance against tumors; therefore, they have reduced cytotoxic properties [9, 10].

6. **In Eye Infection:** Colostrum acts as an effective inhibitor in eye infection against the bacteria, i.e., *Chlamydsia tracomatis*. Even colostrum is topically applied it effectively acts against dryness of eye and eye injury [2].

7. **In Prevention of NSAID Induced Gut Damage:** Researcher's utilized *in vivo* and *in vitro* models to reduce gastrointestinal damage occur by consuming non-steroidal anti-inflammatory agent.

   Colostrum helps in proliferation and cell migration. It was established that bovine colostrum could provide a novel, economical approach for the prevention and treatment of the adverse effect of NSAID'S on the gut, might be value for the treatment of other ulcerative complaint of bowel [3–6].

8. **Antiviral and Antibacterial Activities:** Lactoferrin is a protein that is constituents of colostrum involved in our body's defenses to the infection, including those caused by various bacteria and viruses. Lactoferrin possesses antiviral and antibacterial activity towards a variety of viruses and bacteria [41].

9. **In Weight Loss Program:** The bovine colostrum supplements are the very effective therapy for weight reduction, especially among diabetic and obese populations. Colostrum contains a constituent's named leptin that induces a feeling of fullness and reduced desired to eat more in addition to IGF-1, which is required for the metabolism of fat and energy production occurring through the Krebs cycle. Leptin also work with IGF-1 to reduce the elevated level of cholesterol and triglyceride [41].

10. **Respiratory Disease:** In dealing with something as elusive as the common cold or as invasive as influenza, the best offense is a good defense. Coupling of nutrition's diet with a program of exercise and routine supplementation with high quality bovine colostrum is the best potential defense. The thymus can be capable of be restored to its normal function by the growth factors in colostrum. In addition, colostrum contains particular hormones that regulate the task of thymus and other substances that help to retain the immune system under organized to respond to possible infections before they develop to established [19, 20, 23, 24].

11. **Acquired Immune Deficiency Syndrome (AIDS):** This is a thought-provoking consequence in the individual infected with the human immunodeficiency virus (HIV) since the immune system is the prime focus of outbreak by the virus. As indicated above; the IGF-1 originate in colostrum has been made known to be accomplished of restoring the thymus to its normal functioning capacity. In addition, colostrum comprises mutually the alpha and beta chains of thymosins, which are hormones that have been publicized to work individually and shows the role of thymus. Other substance create in colostrum may in addition of use 1 to AIDS patient. In several studies a lactoferrin, element of colostrum has been made to known to prevent infection with definite viruses to which the AIDS enduring may turn into susceptible. Antibodies existing in extreme superiority bovine colostrum boast too fixed in adjunct efficient in portion to regulate the unadorned diarrhea commonly coupled with the opportunistic enteric infection experienced by few AIDS patients [12, 19, 25, 26].

12. **Leaky Gut Syndrome:** Leaky gut syndrome is related with many autoimmune diseases such as diabetes, inflammatory and irritable bowel disease, chronic fatigue syndrome, and multiple sclerosis. Antibodies formed by the biological system in response to stressors can get attached to tissues through the body and cause inflammation, progressively generated autoantibodies result in chronic inflammatory disorder. The immune system enhancer present in colostrum has revealed markedly beneficial effects on clinical as well as subclinical infection and chronic pain disorder. Thus BC supplements may provide GI and immunological value and help to develop gut mucosal integrity and immunological status [10].

## 1.4   CONCLUSION

The bovine colostrum is utilized as a food supplement which cures a wide variety of diseases. The primary s advantage of this food supplement is that they have very insignificant side effects, and it is well accepted. Bovine colostrum contains various antibodies like IgA, IgM, IgG, which are used to strengthen our immune system to fight against bacteria and viruses. Nowadays, society more interested in bovine colostrum food supplements because they promote immunity, fight infection, and improve gut health.

This bovine colostrum food supplement remain indicative of future predictions for helping in mitigating the diseases such as AIDS, cardiovascular disease, respiratory disease, GI disorder, infectious disease wound healing, and certain cancer.

## KEYWORDS

- acquired immune deficiency syndrome
- bovine
- colostrum
- cyclic adenosine monophosphate
- high-density lipoprotein
- human immunodeficiency virus
- insulin-like growth factor
- low density lipoprotein

## REFERENCES

1. Wilson, J., (1997). Immune system breakthrough: Colostrum. *J. Longevity Res., 3*, 7–10.
2. Chaumeil, C., Loitet, S., & Kogbe, O., (1994). Treatment of severe eye dryness and problematic eye lesions with enriched bovine colostrum lactoserum. *Adv. Exp. Med. Biol., 350*, 595–599.
3. Playford, R. J., Floyd, D. N., Macdonald, C. E., Calnan, D. P., Adenekan, R. O., Johnson, W., Goodland, R. A., & Marchant, T., (1999). Bovine colostrum is a health food supplement which prevents NSAID induced gut damage. *Gut., 44*, 653–658.
4. Simmen, F. A., Cera, K. R., & Mahan, D. C., (1990). Stimulation by colostrum or mature milk of gastrointestinal tissue development in newborn pigs. *J. Anim. Sci., 68*, 3596–3603.
5. Playford, R. J., Watanaba, P., & Woodman, A. C., (1993). Effect of luminal growth factor preservation on intestinal growth. *Lancet, 341*, 843–848.
6. Svanes, K., Itoh, S., & Takeuchi, K., (1982). Restitution of the surface epithelium of the *in vitro* frog gastric mucosa after damage with hyper molar sodium chloride. *Gastroenterology, 82*, 1409–1426.
7. Ginjala, V., & Pakkanen, R., (1998). Determination of transforming growth factor-beta 1 and 2 insulin-like growth factor in bovine colostrum samples. *J. Immunoassay, 19*, 195–207.
8. Tollefsen, S. E., Lajara, R., McCusker, R. H., Clemmons, D. R., & Rotwein, P., (1989). Insulin-like growth factors in muscle development. *J. Biol. Chem., 264*, 13810–13817.

9. Hammon, H. M., & Blum, J. W., (2002). Feeding different amounts of colostrum or only milk replacer modifies receptors of intestinal insulin-like growth factors and insulin in calves. *Domest. Anim. Endocrinol., 22*(3), 155–168.

10. Hurley, W. L., (2000). *Animal Science 308 (on-line)* (p. 11). The Neonate and Colostrum, University of Illinois Urbana-Champagne.

11. Lawrence, H. S., & Borkowsky, W., (1996). Transfer factor-current status and future prospects. *Biotherapy, 9*(1–3), 1–5.

12. Lonnerdal, B., & Iyer, S., (1995). Lactoferrin: Molecular structure and biological function. *Ann. Rev. Nutrition, 13*, 93–110.

13. Cameron, C. M., et al., (1988). The acute effects of growth hormone on amino acid transport and protein synthesis are due to its insulin in-like action. *Endocrinol., 122*(2), 471–474.

14. Shing, Y., & Elagabrun, M., (1987). Purification and characteristic s of a bovine colostrum-derived growth factor. *Molec. Endocrinol., 25*(3), 335–340.

15. Hwa, V., et al., (1999). The insulin-like growth factor binding protein (IGFBP) superfamily. *Endocrin. Rev., 20*(6), 761–787.

16. LeRoith, D., (1996). Insulin-like growth factor receptors and binding proteins. *Clin, Endocrinol. Metab., 10*(1), 49–73.

17. Baratta, M., (2002). Leptin-from a signal of adiposity to a hormonal mediator in peripheral tissues. *Med. Sci. Monit., 8*(12), 282–292.

18. Guerre-Millo, M., (2002). Adipose tissue hormones. *J. Endocrinol. Invest., 25*(10), 855–861.

19. Bjorback, C., & Hollenberg, A. N., (2002). Leptin and melanocortin signaling in the hypothalamus. *Vita. Horm., 65*, 281–311.

20. Kelly, K. M., Oh, Y., Gargosky, S. E., Gucev, Z., Matsumoto, T., Hwa, V., Ng, L., Simpson, D. M., & Rosenfeld, R. G., (1996). Insulin-like growth factor-binding proteins (IGFBPs) and their regulatory dynamics. *Int. J. Biochem. Cell Biol., 28*(6), 619–637.

21. Andrew, D., & Aspinall, R., (2002). Age-associated thymic atrophy is linked to a decline in IL-7 production. *Exp. Gerontol., 37*(2/3), 455–463.

22. Aspinall, R., et al., (2002). Age-associated changes in thymopoesis. *Springer Semin. Immunopathol., 24*(1), 87–101.

23. Fry, T. J., & Mackall, C. L., (2002). Current concepts of thymic aging. *Springer Semin Immunopathol., 24*(1), 7–22.

24. He, F., et al., (2001). Modulation of human humoral immune response through orally administered bovine colostrum. *FEMS Immunol. and Med. Microbiol., 31*, 93–96.

25. Steijns, J. M., & Van, H. A. C., (2000). Occurrence, structure, biochemical properties, and technological characteristics of lactoferrin. *Brit. J. Nutr., 84*(1), 11–17.

26. Moddoveanu, Z., (1983). Antibacterial properties of milk: IgA, peroxidase lactoferrin interactions. *Ann. N.Y. Acad. Sci., 409*, 848–850.

27. Bjornvad, C. R., Thymann, T., Deutz, N. E., et al., (2008). Enteral feeding induces diet-dependent mucosal dysfunction, bacterial proliferation, and necrotizing enterocolitis in preterm pigs on parenteral nutrition. *Am. J. Physiol. Gastrointest Liver Physiol., 295*, G1092–G1103.

28. Jensen, M. L., Sangild, P. T., Lykke, M., et al., (2013). Similar efficacy of human banked milk and bovine colostrum to decrease incidence of necrotizing enterocolitis in preterm piglets. *Am. J. Physiol. Regul. Integr. Comp. Physiol., 305*, R4–R12.

29. Elfstrand, L., & Flóren, C. H., (2010). Management of chronic diarrhea in HIV-infected patients: Current treatment options, challenges and future directions. *HIV Aids (Auckl.), 2*, 219–224.
30. Saxon, A., & Weinstein, W., (1987). Oral administration of bovine colostrum anticryptosporidia antibody fails to alter the course of human cryptosporidiosis. *J. Parasitol., 73*, 413–415.
31. Rump, J. A., Arndt, R., Arnold, A., et al., (1992). Treatment of diarrhea in human immunodeficiency virus-infected patients with immunoglobulins from bovine colostrum. *Clin. Investig., 70*, 588–594.
32. Plettenberg, A., Stoehr, A., Stellbrink, H. J., et al., (1993). A preparation from bovine colostrum in the treatment of HIV-positive patients with chronic diarrhea. *Clin. Investig., 71*, 42–45.
33. Flóren, C. H., Chinenye, S., Elfstrand, L., et al., (2006). Colo plus, a new product based on bovine colostrum, alleviates HIV-associated diarrhea. *Scand. J. Gastroenterol., 41*, 682–686.
34. Kaducu, F. O., Okia, S. A., Upenytho, G., et al., (2011). Effect of bovine colostrum-based food supplement in the treatment of HIV-associated diarrhea in Northern Uganda: A randomized controlled trial. *Indian J. Gastroenterol., 30*, 270–276.
35. Stephan, W., Dichtelmuller, H., & Lissner, R., (1990). Antibodies from colostrum in oral immunotherapy. *J. Clin. Chem. Clin. Biochem., 28*, 19–23.
36. Feasey, N. A., Healey, P., & Gordon, M. A., (2011). Review article: The etiology, investigation and management of diarrhea in the HIV-positive patient. *Aliment Pharmacol. Ther., 34*, 587–603.
37. Seth, R., & Das, A., (2011). Colostrum powder and its health benefits. In: Sharma, R., & Mann, B., (eds.), *Chemical Analysis of Value-Added Dairy Products and Their Quality Assurance* (pp. 59–67). Karnal, Haryana, India: Division of Dairy Chemistry, National Dairy Research Institute.
38. Macy, I. G., (1949). Composition of human colostrum and milk. *Am. J. Dis. Child., 78*, 589–603.
39. Moddoveanu, Z., (1983). Antibacterial properties of milk: IgA, peroxidase lactoferrin interactions. *Ann. N.Y. Acad. Sci., 409*, 848–850.
40. Anwar, A., Gaspz, J. M., Pampallona, S., Zahid, A. A., Sigaud, P., Pichard, C., & Brink, M., (2002). Effect of congestive heart failure on the insulin-like growth fac tor-1 system. *Am. J. Cardiol., 90*(12), 1402–1405.
41. Lissner, R., et al., (1996). A standard immunoglobulin preparation produced from bovine colostrum shows antibody reactivity and neutralization activity against Shiga-like toxins and EHEC-hemolysis in of *Escherichia coli* O157:h7. *Infection, 24*(5), 378–383.
42. Chelack, B. J., Morley, P. S., & Haines, D. B., (1993). Evaluation of methods for dehydration of bovine colostrum for total replacement of normal colostrum in calves. *Can. Vet. J., 34*(7), 407–412.
43. Godden, S. M., Smith, S., Feirtag, J. M., Green, L. R., Wells, S. J., & Fetrow, J. P., (2003). Effect of on-farm commercial batch pasteurization of colostrum on colostrum and serum immunoglobulin concentrations in commercial dairy calves. *J. Dairy Sci., 86*(4), 1503–1512.
44. Manohar, A., Williamson, M., & Koppikar, G. V., (1997). Effect of storage of colostrum in various containers. *Indian Pediatrics, 34*(4), 293–295.
45. McGuirk, S. M., (1998). Colostrum, quality and quantity. *Cattle Practice, 6*, 63–66.

46. Moore, M., Tyler, J. W., Munase, C., Dawes, M., & Middleton, J. R., (2005). Effect of delayed colostrum collection on colostral IgG concentration in dairy cows. *JAVMA, 226*(8), 1375–1377.

47. Hardy, G., (2000). Nutraceuticals and functional foods: Introduction and meaning. *Nutr., 16*(7/8), 688, 689.

48. McConnell, M. A., Buchan, G., Borissenko, M. V., & Brooks, H. J. L., (2001). A comparison of IgG and IgG1 activity in an early milk concentrate from non-immunized cows and a milk from hyperimmunized animals. *Food Research International, 34*(2/3), 255–261.

# CHAPTER 2

# Molecular Modeling: Novel Techniques in Food and Nutrition Development

KHEMCHAND R. SURANA,[1] EKNATH D. AHIRE,[1]
VIJAYRAJ N. SONAWANE,[1] SWATI G. TALELE,[2] and G. S. TALELE[3]

[1]Divine College of Pharmacy, Nampur Road, Satana, Maharashtra, India

[2]Sandip Institute of Pharmaceutical Sciences, Mahiravani, Nasik, Maharashtra, India, E-mail: swatitalele77@gmail.com

[3]Matoshri College of Pharmacy, Eklhre Nasik, Maharashtra, India

## ABSTRACT

Molecular modeling is central for understanding chemistry by providing improved tools for investigating, interpreting, explaining, and discovering new phenomena. Molecular modeling is easy to perform with presently obtainable software, but the difficulty lies in getting the right model and appropriate interpretation. There are two foremost modeling methodologies are used for the conception of new nutrition, which are direct nutrition design and indirect nutrition design and also having dissimilar methodologies for computing the energy of the molecule. First one is quantum mechanics, which comprises ab initio, DFT, and semi-empirical methodologies, and the other one is the molecular mechanics. Energy minimization methods can be separated into diverse classes depending on the order of the derivative used for pinpointing a minimum on the energy surface. The most significant concerns in medicinal chemistry and food and nutrition research are structure elucidation, conformational analysis, physicochemical characterization, and biological activity determination. The persistence of molecular structure is significant as the structure of the molecule predicts the physical, chemical, and biological belongings of the molecule. The main objective of conformational analysis is to gain insight on conformational characteristic of flexible biomolecules and nutrition but to also recognize the relation amongst the role of conformational tractability and their activity.

## 2.1    INTRODUCTION

Modeling is a tool for doing chemistry. Models are center for learning of chemistry. Molecular modeling permits us to do and clarify chemistry healthier by giving good tools for investigating, interpreting, explaining, and ascertaining new phenomena. Like investigational chemistry, it is a skill-demanding science and should be learnt via doing and not just reading. Molecular modeling is easy to execute with presently accessible software, but the trouble lies in receiving the right model and appropriate interpretation [1–3].

### 2.1.1    MOLECULAR MODELING TOOLS

The tools of the trade have progressively evolved from physical models and calculators, comprising the use of programmable calculators, computers as visualization aids, computers running commercially written analysis packages such as SYBYL and furthermost lately integration by means of internet grounded tools and workbenches founded on HTML, JavaScript, etc. The following tools are obligatory for modeling of nutrition by means of computers:

1.  **Hardware:** Various classes of computers are essential for molecular modeling. For chemical information systems, the selection of a computer is generally larger, and a lot of packages run on VAX, IBM, or PRIME machines. Presently, the molecular modeling community is by means of equipment from manufacturers like, Digital, IBM, Sun, Hewlett-Packard, and Silicon Graphics running with the UNIX operating system.
2.  **Software Components:** A multiplicity of commercial packages is existing for PC-based systems as well as supercomputer grounded systems. Presently, few of the molecular modeling software that is existing for commercial and academic molecular modeling functions such as MOE, Chemsketch, AMBER, Catalyst, etc.

The computational chemistry programs permit scientists to produce and present molecular data comprising geometries (bond lengths, bond angles, and torsion angles), energies (heat of formation, activation energy, etc.), and characteristics (volumes, surface areas, diffusion, viscosity, etc.), [2, 4].

### 2.1.2   MOLECULAR MODELING STRATEGIES

Presently, two foremost modeling approaches are applied for the commencement of new nutrition [5]:

1. **Direct Nutrition Design:** In the direct methodology, the three-dimensional properties of the well-known receptor site are determined from X-ray crystallography to scheme a lead molecule. In direct design, the receptor site geometry is well known; the difficulty is to discover a compound that fulfills some formal constraints and is also a better chemical match. After verdict good candidates rendering to these benchmarks, a docking step with energy minimization can be applied to estimate binding strength [6].
2. **Indirect Nutrition Design:** This methodology includes comparative analysis of structural properties of known active and inactive compounds that are complementary with a hypothetical receptor site. If the site geometry is not known, as is often the case, the developer must base the design on other ligand compounds that bind well to the site [7, 8].

## 2.2   QUANTUM MECHANICS AND MOLECULAR MECHANICS

There are two different approaches to compute the energy of a molecule (Figure 2.1) [9]:

1. Quantum mechanics; and
2. Molecular mechanics.

### 2.2.1   QUANTUM MECHANICS

A technique based on first principles. In this methodology, nuclei are organized in the space, and the analogous electrons are extent all over the system in a nonstop electronic density and computed by resolving the Schrödinger equation. For biomolecules, this procedure can be done inside the Born-Oppenheimer approximation, and for the greatest of the purposes, the Hartree-Fock self-consistent field is the utmost suitable process to compute the electronic density and the energy of the organization. When chemical reactions do not requirement to be simulated, classical mechanics can designate the performance of a biomolecular system. This mathematical model is recognized as molecular mechanics, and can be applied to compute the energy of systems comprising an enormous number of atoms, such as compounds or complex systems of biochemical and biomedical importance [10, 11].

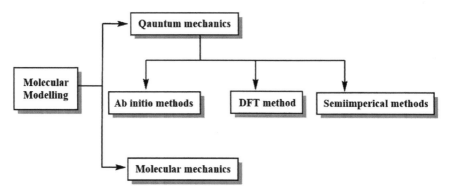

**FIGURE 2.1**    Different approaches of molecular modeling.

Quantum mechanics is fundamentally the molecular orbital calculation and provides the utmost exhaustive description of a molecule's chemical performance. Highest energy occupied molecular orbital (HOMO) and lowest energy unoccupied molecular orbital (LUMO), quantum approaches employ the principles of particle physics to inspect structure as a utility of electron dissemination. Geometries and properties for transition state and excited state will be calculated with Quantum mechanics. Their applications can be prolonged to the analysis of compounds as yet un-synthesized and chemical species which are problematic (or impossible) to separate [12].

Quantum mechanics is based on the Schrödinger equation [13]:

$$H\varphi = E\varphi = (U + K)\varphi$$

where; $\varphi$ = wave function describes the electron distribution around the molecule; U = potential energy; K = kinetic energy; E = energy of the system relative to one in which all atomic particles are separated to infinite distances; H = Hamiltonian for the system. It is an "operator," a mathematical construct that operates on the molecular orbital, $\varphi$ to determine the energy.

### 2.2.1.1  AB INITIO METHODS

The word Ab initio is Latin for "from the beginning" grounds of quantum theory. This is a rough quantum mechanical calculation for a utility or finding an imprecise solution to a discrepancy equation. In its unpolluted form, quantum theory applied well-known physical constants such as the velocity of light, values for the masses and charges of nuclear particles, and differential equations to straight calculate molecular characteristics and geometries. This formalism is mentioned to as ab initio (from first principles) quantum mechanics [14, 15].

*Hartree $\pm$ Fock Approximation*

The utmost common type of Ab initio calculation in which the primary approximation is the central field approximation means Coulombic electron-electron repulsion is taken into account by integrating the repulsion term. This is a variational calculation, means that the approximate energies calculated are all equivalent to or greater than the exact energy. The energies are calculated in units known as Hartrees (1 Hartree = 27.2116 eV) [16].

The steps in a Hartreefock calculation initiated with an initial guess for the orbital coefficients, frequently by means of a semi-empirical technique. This utility is used to calculate energy and a novel set of orbital coefficients, which can then be used to gain a new set, and so on. This technique remains iteratively until the energies and orbital coefficient remains constant from one iteration to the next. This iterative technique is called a self-consistent field process (SCF). It breaks the many-electron Schrodinger equation into numerous simpler one-electron equations. Each one-electron equation is resolved to yield a single-electron wave function, named an orbital, and energy, termed an orbital energy. In different of quantum mechanics, molecular mechanics disregard electrons and compute the energy of a system only as significance of the nuclear locations. Then, it is promising to take into account in an unspoken way the electronic component of the system by satisfactory parameterization of the potential energy function. The set of equations and constraints which define the potential surface of a molecule is named *force field* [15, 17].

### 2.2.1.2   DENSITY FUNCTIONAL THEORY (DFT)

How somewhat (s) is dispersed/spread about a given space. Electron density tells us where the electrons are likely to exist [18].

$$p \propto \varphi * \varphi$$

A function depends on a set of variables.

$$y = f(x)$$

For example, wave function depends on electron coordinates.
*What is a functional?* A functional depends on functions, which in turn depends on a set of variables.

$$E = F[F(x)]$$

For example, energy depends on the wave function, which depends on electron coordinates.

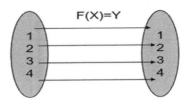

The electron density is the square of wave role/function and incorporated over electron coordinates. The difficulty of a wave function rises as the number of electrons grows up, however, the electron density still depends only on three coordinates. Through this theory, the stuffs of a many-electron system can be determined by means of functional, i.e., roles of alternative function, which in this case are the spatially reliant electron density [19].

There are problems in by means of density functional theory (DFT) to correctly designate intermolecular interactions, particularly van der Waals forces (dispersion); charge transfer excitations; transition states, overall potential energy surfaces and roundabout other intensely correlated systems [20].

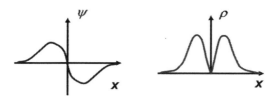

### 2.2.2   MOLECULAR MECHANICS

The electrons and nuclei of the atoms are not clearly comprised in the calculations. Molecular mechanics deliberates a molecule to be a collection of masses relating one with each other via harmonic forces. Therefore, the atoms in molecules are behaved as ball of dissimilar sizes and flavors combined together by springs of variable strength and equilibrium distances (bonds). This generalization permits by means of molecular mechanics as a fast computational model that can be used to molecules of any size [21].

In the path of a calculation the total energy is diminished with related to the atomic coordinates, and comprises in a sum $_{of}$ dissimilar aids that compute the deviations since equilibrium of bond lengths, angles, and torsions plus non-bonded interactions [22]:

$$E_{tot} = E_{str} + E_{bend} + E_{tors} + E_{vdw} + E_{elec} + \ldots$$

where; $E_{str}$ is the bond-stretching energy term; $E_{tot}$ is the total energy of the molecule; $E_{vdw}$ is the van der Waals energy term; $E_{elec}$ is the electrostatic energy term; $E_{tors}$ is the torsional energy term; $E_{bend}$ is the angle-bending energy term.

The first term in the equation designates the energy alteration as a bond stretches and contracts from its ideal unstrained length. It is expected that the inter-atomic forces are harmonic, so the bond-stretching energy term can be designated by a simple quadratic function shown as the following equation:

$$E_{str} = \frac{1}{2} k_b \left( b - b_0 \right)^2$$

where; $k_b$ is the bond-stretching force constant, $b_0$ is the unstrained bond length, and $b$ is the genuine bond length:

$$E_{bend} = \frac{1}{2} k_\theta \left( \theta - \theta_0 \right)^2$$

where; $k_q$ is the angle-bending force constant, $\Theta_0$ is the equilibrium value for the bond angle $\Theta$, and $\Theta$ is the real value for $\Theta$. Correspondingly for angle bonding a simple harmonic, spring-like demonstration is engaged. The presence labeling the angle-bending term is given in the above equation.

A communal expression for the dihedral potential energy term is a cosine series as given in the following equation:

$$E_{tors} = \frac{1}{2} k_j \left( 1 + \cos \left( n\varphi - \varphi_0 \right) \right)$$

where; $k_j$ is the torsional barrier, $\phi$ is the actual torsional angle, $n$ is the periodicity (number of energy minima through a full cycle), and $\phi_0$ is the reference torsional angle.

The van der Waals connections between not straight connected atoms are typically characterized by a Lennard-Jones potential shown in the following equation:

$$E_{vdw} = \sum \frac{A_{ij}}{r_{ij}^{12}} - \frac{B_{ij}}{r_{ij}^{6}}$$

where; $A_{ij}$ is the repulsive term coefficient. $B_{ij}$ is the attractive term coefficient and $rij$ is the distance between the atoms $i$ and $j$. In order to designate the electrostatic forces a surplus term with the Coulomb interaction is applied as equation:

$$E_{elec} = \frac{1}{\varepsilon} \frac{Q_1 Q_2}{r_{ij}}$$

where; $\varepsilon$ is the dielectric constant, and $Q_1$ and $Q_2$ are atomic charges of relating atoms and $r_{ij}$ is the inter-atomic distance.

The equilibrium values of these bond lengths and bond angles are the conforming force constants applied in the potential energy function definite in the force field and define a set known as *force field parameters*. Respectively, deviation from these equilibrium values will outcome in increasing the total energy of the molecule. Therefore, the total energy is a measure of intramolecular strain comparative to a hypothetical molecule with an ultimate geometry of equilibrium. Through itself, the total energy has no strict physical significance, but alterations in total energy between two different conformations of the similar molecule can be matched [23–25].

## 2.3    ENERGY-MINIMIZING METHODS

Energy minimization approaches can be separated into diverse classes depending on the order of the derivative applied for tracing a lowest on the energy external. Zero-order approaches are those that only used the energy function to recognize regions of low energy via a grid search process. The utmost well-known technique of this kind is the SIMPLEX technique. Inside first-derivative methods, there are numerous measures like the steepest descent technique or the conjugate gradient technique that mark the use of the gradient of the function. Second-derivative approaches, like the Newton-Raphson algorithm, mark usage of the hessian to locate minima [26, 27].

### 2.3.1    STEEPEST DESCENT METHOD

In the steepest descent technique, the minimizer computes statistically the first derivative of the energy purpose to discover a minimum. The energy is calculated for the preliminary geometry and then once more after one of the atoms has been relocated in a small increment in one of the directions of the coordinate scheme. This progression is frequent for all atoms which lastly are relocated to a new position downhill on the energy surface. The technique stops when a programed threshold condition is satisfied. The steepest descent technique is frequently use for structures far from the minimum as a first, rough, and introductory run marked by a successive minimization commissioning a more progressive algorithm like the conjugate gradient [28–30].

### 2.3.2    CONJUGATE GRADIENT METHOD

The conjugate gradient algorithm accumulates the evidence related to the role from one repetition to the next. With this continuing the reverse of the development prepared in an earlier iteration can be avoided. For to each minimization step, the gradient is calculated and applied as additional information for computing the new way vector of the minimization technique. Consequently, each succeeding step improves the direction towards the minimum. The computational exertion and the storage necessities are superior to for steepest descent; however, conjugate gradients are the technique of choice for greater systems. The superior total computational expenditure and the lengthier time per repetition is more than compensated by the supplementary well-organized meeting to the minimum attained by conjugate gradients. As a summary, the choice of the minimization method depends on two parameters: the size of the system and the existing state of the optimization. For structures far from minimum, as a universal rule, the steepest descent technique is often the finest minimizer to use for the 100–1000 repetitions. The minimization can be accomplished to merging with conjugate gradients [31–33].

There are numerous techniques in molecular minimization to describe convergence criteria. In non-gradient minimizers, only the upsurges in the energy and/or organization can be taken to judge the quality of the definite geometry of the molecular system. In all gradient minimizers, nevertheless, atomic gradients are applied for this persistence. The utmost technique in this esteem is to calculate the root mean square gradients of the forces on individual atom of a molecule. The value selected as a maximum derivative will ground on the objective of the minimization. If a simple relaxation of a strained molecule is wanted, a rough convergence benchmark like a maximum derivative of 0.1 kcal $mol^{-1}Å^{-1}$ is adequate, whereas for other cases, convergence to a maximum derivative less than 0.001 kcal $mol^{-1}Å^{-1}$ is essential to discover a final minimum [32, 34].

## 2.4   CONFORMATIONAL ANALYSIS

The utmost significant fears in medicinal chemistry and food and nutrition research are structure elucidation, conformational analysis, physicochemical characterization, and biological activity determination. The determination of molecular structure is important as the structure of the molecule expects the physical, chemical, and biological properties of the molecule. Conformational hunt approaches find uses in the design of targeted chemical hosts and nutrition discovery. Conformations are diverse 3D spatial engagements of the atoms in particles are inter-convertible by free revolution of single bonds [35].

The main objective of the conformational investigation is to advance insight on conformational distinguishing of flexible biomolecules and nutrition; however, to also recognize the relation amongst the role of conformational elasticity and their activity. Consequently, it shows an important role in computer-aided design too. The importance of conformational analysis not just spreads to computational docking and screening, however, also for lead optimization. DHR Barton is deliberated the most significant contributor to modern conformational analysis. In 1950, he presented how different substituents at the equatorial and axial positions affect the rate of reactivity of substituted cyclohexane. Recognition of all promising minimum energy structures (conformations) of compounds is the goal of conformational analysis [36, 37].

Conformational analysis is a computational technique in which restraints are applied such that the molecule believes a conformation alike to the inflexible template molecule. Conformational analysis is a challenging problem since even simple compounds may have a great number of conformational isomers. The typical approach in conformational analysis is to application a search algorithm to produce a series of initial conformations. Respectively of these in turn are then exposed to energy minimization in order to derive the related minimum energy structure. Global minimum-energy conformation is the conformation with the lowermost energy. It is not authoritative that the global energy minimum conformation is the bioactive conformation of the nutrition. Maximum of nutrition being flexible compounds can used of alterations and rotations about rotatable bonds assume a large number of conformations. Pharmacophore is a collection of steric and electronic properties that are vital to safeguard optimal communication amongst the precise biological targets (receptor/enzyme) so as to prohibit a biological response [37–39].

The furthermost difficult and main problem in 3D pharmacophore acknowledgment is the identification of the receptor-bound/bioactive conformation. This can be accomplished by studies of the spatial prearrangement of the bioactive conformer so as to express the 3D pharmacophore. This conformation could be the local minimum, global minimum, or any transition state amongst the local minima. Even though it is common practice to adopt that global minimum is the bioactive conformation [37, 40].

## 2.5    GLOBAL CONFORMATIONAL MINIMA DETERMINATION

In the conformational search stage, structural factors of the ligands, such as torsional (dihedral), translational, and rotational degrees of freedom, are incrementally adapted. Conformational search procedures perform this task by relating organized and stochastic search approaches. Methodical search approaches encourage slight differences in the structural factors, progressively altering the conformation of the ligands. The algorithm probes the energy landscape of the conformational space and, after abundant search and assessment cycles, converges to the minimum energy solution conforming to the utmost likely binding mode. Even though the technique is operative in sightseeing the conformational space, it can touch to a local minimum rather than the global minimum [41, 42].

This disadvantage can be overwhelmed by acting simultaneous searches starting from diverse points of the energy landscape (i.e., distinct conformations). Stochastic approaches bring out the conformational search by arbitrarily altering the structural factors of the ligands. For this, the algorithm produces ensembles of molecular conformations and populates an extensive range of the energy landscape [43]. This approach evades trapping the final solution at a local energy minimum and enhancing the possibility of finding a global minimum. As the algorithm encourages an extensive coverage of the energy landscape, the computational cost related to this process is a significant limitation [44].

## 2.6    CONCLUSION

Molecular modeling is easy to perform with presently existing software; however, the trouble lies in receiving the right model and correct interpretation. The apparatuses of the trade have progressively evolved from physical models and calculators, comprising the usage of programmable calculators, computers as visualization aids, computers running commercially written analysis packages such as SYBYL and utmost lately integration by means of internet grounded tools and workbenches founded on HTML, JavaScript, etc. The molecular modeling community is by means of equipment from manufacturers such as Digital, IBM, Sun, Hewlett-Packard, and Silicon Graphics running with the UNIX operating system and some of the molecular modeling software that are existing for commercial and academic molecular modeling meanings such as MOE, Chemsketch, AMBER, Catalyst, etc. The choice of the minimization technique rest on two factors: the size of the system and the existing state of the optimization. For structures far from minimum, as an overall rule, the steepest descent technique is often the best minimizer to usage for the 100–1000 iterations. The minimization can be finished to convergence with conjugate gradients. Conformational search approaches find applications in the design of targeted chemical hosts and nutrition discovery. Conformations are different 3D spatial engagements of the atoms in a molecule are inter-convertible by free revolution of single bonds.

## KEYWORDS

- **density functional theory**
- **highest energy occupied molecular orbital**
- **lowest-energy unoccupied molecular orbital**
- **minimization**
- **molecular modeling**
- **self-consistent field process**

## REFERENCES

1. Schlick, T., (2010). *Molecular Modeling and Simulation: An Interdisciplinary Guide: An Interdisciplinary Guide* (Vol. 21). Springer Science & Business Media.
2. Jensen, J. H., (2010). *Molecular Modeling Basics*. CRC Press.
3. Cygan, R. T., & Kubicki, J. D., (2018). *Molecular Modeling Theory: Applications in the Geosciences* (Vol. 42). Walter de Gruyter GmbH & Co KG.
4. Gubbins, K. E., & Moore, J. D., (2010). Molecular modeling of matter: Impact and prospects in engineering. *Industrial and Engineering Chemistry Research, 49*(7), 3026–3046.
5. Amann, A., Boeyens, J. C., & Gans, W., (2013). *Fundamental Principles of Molecular Modeling*. Springer Science & Business Media.
6. Ferreira, L. G., Oliva, G., & Andricopulo, A. D., (2015). Target-based molecular modeling strategies for schistosomiasis nutrition discovery. *Future Medicinal Chemistry, 7*(6), 753–764.
7. Sun, A. Y., et al., (2013). Modeling the impact of alternative strategies for rapid molecular diagnosis of tuberculosis in Southeast Asia. *American Journal of Epidemiology, 178*(12), 1740–1749.
8. Ferreira, L. G., et al., (2015). Molecular docking and structure-based nutrition design strategies. *Molecules, 20*(7), 13384–13421.

9. Atkins, P. W., & Friedman, R. S., (2011). *Molecular Quantum Mechanics*. Oxford University Press.
10. Levine, R. D., (2011). *Quantum Mechanics of Molecular Rate Processes*. Courier Corporation.
11. Wallrapp, F. H., & Guallar, V., (2011). Mixed quantum mechanics and molecular mechanics methods: Looking inside proteins. *Wiley Interdisciplinary Reviews: Computational Molecular Science, 1*(2), 315–322.
12. Merkt, F., & Quack, M., (2011). Molecular quantum mechanics and molecular spectra, molecular symmetry, and interaction of matter with radiation. *Handbook of High-Resolution Spectroscopy*.
13. Longhi, S., (2015). Fractional Schrödinger equation in optics. *Optics Letters, 40*(6), 1117–1120.
14. Ohno, K., Esfarjani, K., & Kawazoe, Y., (2018). *Computational Materials Science: From Ab Initio to Monte Carlo Methods*. Springer.
15. Carsky, P., & Urban, M., (2012). *Ab Initio Calculations: Methods and Applications in Chemistry, 16*. Springer Science & Business Media.
16. Fuentes, M. J., et al., (2012). Structural, electronic and acid/base properties of [Ru (bpy (OH) 2) 3] 2+(bpy (OH) 2= 4, 4'-dihydroxy-2, 2'-bipyridine). *Dalton Transactions, 41*(40), 12514–12523.
17. Schaefer, III. H. F., (2012). *Quantum Chemistry: The Development of Ab Initio Methods in Molecular Electronic Structure Theory*. Courier Corporation.
18. Obot, I., Macdonald, D., & Gasem, Z., (2015). Density functional theory (DFT) as a powerful tool for designing new organic corrosion inhibitors. Part 1: An overview. *Corrosion Science, 99*, 1–30.
19. Labanowski, J. K., & Andzelm, J. W., (2012). *Density Functional Methods in Chemistry*. Springer Science & Business Media.
20. Burke, K., (2012). Perspective on density functional theory. *The Journal of Chemical Physics, 136*(15), 150901.
21. Van, D. K. M. W., & Mulholland, A. J., (2013). Combined quantum mechanics/ molecular mechanics (QM/MM) methods in computational enzymology. *Biochemistry, 52*(16), 2708–2728.
22. Tarefder, R. A., Pan, J., & Hossain, M. I., (2016). Molecular dynamics simulation of asphaltic material: Molecular dynamics simulations of oxidative aging of asphalt molecules under stress and moisture. In: *Handbook of Research on Advanced Computational Techniques for simulation-Based Engineering* (pp. 334–363). IGI Global.
23. Homeyer, N., & Gohlke, H., (2012). Free energy calculations by the molecular mechanics Poisson-Boltzmann surface area method. *Molecular Informatics, 31*(2), 114–122.
24. Yesselman, J. D., et al., (2012). MATCH: An atom-typing toolset for molecular mechanics force fields. *Journal of Computational Chemistry, 33*(2), 189–202.
25. Chylek, L. A., et al., (2014). Rule-based modeling: A computational approach for studying biomolecular site dynamics in cell signaling systems. *Wiley Interdisciplinary Reviews: Systems Biology and Medicine, 6*(1), 13–36.
26. Wan, W. L., Chan, T. F., & Smith, B., (1999). An energy-minimizing interpolation for robust multigrid methods. *SIAM Journal on Scientific Computing, 21*(4), 1632–1649.
27. Hofer, M., & Pottmann, H., (2004). Energy-minimizing splines in manifolds. In: *ACM Transactions on Graphics (TOG)*. ACM.

28. Fliege, J., & Svaiter, B. F., (2000). Steepest descent methods for multicriteria optimization. *Mathematical Methods of Operations Research, 51*(3), 479–494.
29. Drummond, L. G., & Svaiter, B. F., (2005). A steepest descent method for vector optimization. *Journal of Computational and Applied Mathematics, 175*(2), 395–414.
30. Yuan, Y. X., (2006). A new step size for the steepest descent method. *Journal of Computational Mathematics,* 149–156.
31. Exl, L., et al., (2019). Preconditioned nonlinear conjugate gradient method for micromagnetic energy minimization. *Computer Physics Communications, 235,* 179–186.
32. Dai, Y. H., (2010). *Nonlinear Conjugate Gradient Methods.* Wiley Encyclopedia of Operations Research and Management Science.
33. Lu, Y., et al., (2010). Molecular mechanism of interaction between norfloxacin and trypsin studied by molecular spectroscopy and modeling. *Spectrochimica Acta Part A: Molecular and Biomolecular Spectroscopy, 75*(1), 261–266.
34. Byröd, M., & Åström, K., (2010). Conjugate gradient bundle adjustment. In: *European Conference on Computer Vision.* Springer.
35. Firouzi, M., & Wilcox, J., (2012). Molecular modeling of carbon dioxide transport and storage in porous carbon-based materials. *Microporous and Mesoporous Materials, 158,* 195–203.
36. Rasmussen, K., (2012). *Potential Energy Functions in Conformational Analysis, 37.* Springer Science & Business Media.
37. Mazzanti, A., & Casarini, D., (2012). Recent trends in conformational analysis. *Wiley Interdisciplinary Reviews: Computational Molecular Science, 2*(4), 613–641.
38. Shasha, C., et al., (2014). Nanopore-based conformational analysis of a viral RNA nutrition target. *ACS Nano, 8*(6), 6425–6430.
39. Tormena, C. F., (2016). Conformational analysis of small molecules: NMR and quantum mechanics calculations. *Progress in Nuclear Magnetic Resonance Spectroscopy, 96,* 73–88.
40. Weigelt, S., et al., (2012). Synthesis and conformational analysis of efrapeptins. *Chemistry-A European Journal, 18*(2), 478–487.
41. Olson, B., & Shehu, A., (2011). Populating local minima in the protein conformational space. In: *2011 IEEE International Conference on Bioinformatics and Biomedicine.* IEEE.
42. Ma, D. L., et al., (2011). Molecular modeling of nutrition-DNA interactions: Virtual screening to structure-based design. *Biochimie, 93*(8), 1252–1266.
43. Skovstrup, S., et al., (2010). Conformational flexibility of chitosan: A molecular modeling study. *Biomacromolecules, 11*(11), 3196–3207.
44. Li, H., Lin, Z., & Luo, Y., (2014). Gas-phase IR spectroscopy of deprotonated amino acids: Global or local minima? *Chemical Physics Letters, 598,* 86–90.

# CHAPTER 3

# Informatics and Methods in Nutrition Design and Development

KHEMCHAND R. SURANA,[1] EKNATH D. AHIRE,[1]
VIJAYRAJ N. SONAWANE,[1] SWATI G. TALELE,[2] and G. S. TALELE[3]

[1]Divine College of Pharmacy, Nampur Road, Satana, Maharashtra, India

[2]Sandip Institute of pharmaceutical Sciences, Mahiravani, Nasik, Maharashtra, India, E-mail: swatitalele77@gmail.com

[3]Matoshri College of Pharmacy, Eklhre Nasik, Maharashtra, India

## ABSTRACT

Bioinformatics is a noteworthy in appreciative the composite mechanisms of the cell and it is very beneficial for biomedical investigators to examination the clinical samples. Bioinformatics includes the usage of methods comprising useful mathematics, computer science, informatics, artificial intelligence (AI), statistics, chemistry, and biochemistry to resolve biological difficulties frequently on the molecular level. Bioinformatics tools provide significant benefits for nutrition discovery agendas such as cost-saving, time-to-market, and in-sight. The practical feature of bioinformatics is to recognize the code of life that is to decode the evidence that exists in the nucleotide sequence. Chemo-informatics is the application of computer and informational methods to challenge chemical problems, which highlight the manipulation of chemical structural information. Bioinformatics having numerous applications such as molecular medicine, personalized medicine, preventive medicine, gene therapy, etc. Several software packages are significant and free, and secure by well-known organizations. These consist of databanks like ChEMBL and SwissSidechain, software tools like 3D visualization, UCSF Chimera, and provides platform for software designers fascinated in structural biology, SwissSimilarity for virtual screening, Swiss-Bioisostere for ligand design, SwissTarget.

## 3.1   INTRODUCTION TO BIOINFORMATICS

The exponential growth in biological data has responsible for the development of primary and secondary databanks of nucleic acid arrangements, protein sequences, and structures are named as bioinformatics. There are some databases available for this development include protein data bank (PDB), GenBank, SWISS-PROT, PIR, CATH, SCOP, and many more, these databanks are widely available online on internet across the world as open access publically [1]. Currently, the elementary investigation is carried by means of these databanks nearby with the support of sequence analysis tools like BLAST, CLUSTALW, FASTA, etc., and the modeled structures are evaluated by MOLMOL, WebLab, Rasmol, etc. The bioinformatics is dignified as significant in thoughtful of the multifarious mechanisms of the cell [2]. Additionally, bioinformatics is extensively useful for biomedical researcher to test the experimental models. The "Clinical Informatics" is the whole process of data gathering to investigation the samples and the results. Bioinformatics comprise the application of methods comprising informatics, applied statistics, mathematics, chemistry, biochemistry, computer science, and artificial intelligence (AI) to resolve biological difficulties typically on the molecular level. Major research focused in the field comprise structure forecast, prediction of gene expression, protein-protein interactions, sequence alignment, gene discovery, genome association, protein structure arrangement and the modeling of development [3–5].

Bioinformatics can affects meaningfully in resolving the following types of complications:

1. To study of structural genomics means prediction of 3D structure grounded on linear genomic information.
2. Forecasting of gene utility and establishing of gene libraries, i.e., gene expression analysis.
3. The arena of proteomics is the capability to usage genome sequence to classify proteins and their purposes, protein interactions, alterations, and functions.
4. Simulation of breakdown from the biochemical utilities of an organism.
5. Prediction of structure from investigational data by molecular modeling and molecular dynamics methods.
6. Food and nutrition designing and detection from well-designed genomics and proteomics data.

Bioinformatics is an important tool in objective discovery and development with human genome sequencing, and the integral part this is *in-silico* analysis of gene expression and gene function, which helps in the assortment of the utmost appropriate targets for an ailment under study. Bioinformatics is the essential hub that unites different disciplines and methodologies. On the backing side of the hub, computational resources, and other software technologies like information technology, information management, and databases, provides the infrastructure for bioinformatics [6]. On the scientific side of the hub, bio-informatics approaches are applied broadly in molecular biology, genomics, proteomics, and in CADD research. Bioinformatics in computer-aided nutrition design as shown in Figure 3.1.

Bioinformatics supports CADD research in the following aspects:

- Virtual high-throughput screening;
- Sequence analysis;
- Similarity searches;
- Food and nutrition lead optimization;
- Physicochemical modeling;
- Nutrition bioavailability and bioactivity.

Food and nutrition discovery and development over bioinformatics to regulate the databases for small molecules that are possible lead compounds, to hunt databases of protein. The practical aspect of bioinformatics is to recognize the code of life that is to decode the information reside in nucleotide sequence. It is a famous fact that DNA is the basic molecule of life that unswervingly manages the ultimate biology of virtually all organisms. The nucleotide sequence establishes the genes, which is definite in terms of proteins. Any differences and mistakes in the nucleotide sequence of the genomic DNA or mutations might lead to the growth of genetic disorders or other metabolic variations [7, 8].

Consequently, investigators/scientists in the period of biotechnology or molecular biology prerequisite to know the natural surroundings of individual genomes of different prokaryotic and eukaryotic creatures. Previously many DNA sequencing developments have been accomplished and many more are in development prominent to enormous amount of biological information. This has supplementary in the development of the science of bioinformatics. Handling such massive information and clarification was not conceivable lacking of bioinformatics. Therefore, bioinformatics can be known as the brain of biotechnology [9].

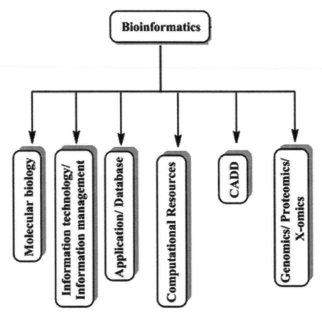

**FIGURE 3.1**    Bio-informatics in computer-aided food and nutrition design.

### 3.1.1   ADVANTAGES OF CADD AND BIOINFORMATICS

CADD methods and bioinformatics tools offer significant benefits for food and nutrition discovery programs [10]:

1. **Cost Saving:** The food and nutrition discovery and development cost has reached $800 million for a single food and nutrition to be successfully brought into the market. Recently, many healthy food and nutrition industries are working on CADD for minimization of investment and cost burden [11].

2. **Time-to-Market:** CADD having potential power to reduce the time period for food and nutrition development and optimization and evades possible 'dead-end' compound on final phase, with the help of this food and nutrition can reach to market more rapidly and cost-effectively [12].

3. **In-Sight:** CADD gives knowledge about nutrition-receptor interaction and atomic-scale binding properties

to particular ligand or protein with the help of molecular graphics methods. This will helps to scientist and give new ideas for modification of the nutrition molecule for improved fit. Consequently, CADD and bioinformatics collected are an influential amalgamation in nutrition investigation and development [10].

## 3.2 BIOINFORMATICS AND NUTRITION DISCOVERY

Nutrition discovery is the step-by-step process in which new nutrition candidates discovered. Traditionally, healthy food and nutrition industries adopted well-established pharmacology and healthy food and nutrition chemistry-based nutrition discovery approaches, and faces various difficulties in finding new nutrition. The most competitions first industry to patent a new chemical entity (NCE) for the treatment of specific disease or disorder takes all the spoils and the other opponents to typically interval for patent terminations to participate in the generosity [10]. Nowadays, therefore, the healthy food and nutrition industries are ready to invest potentially in all those methodologies that display probable to quicken any level of the nutrition development process. Bioinformatics generates remarkable interest generate more and more nutrition candidates in a short period of time with low risk and decreases the stress for the discovery of novel nutrition candidates. Consequently, nowadays being of novel, distinct field for discovery and development of novel nutrition candidates through negligible investment and less time period, which is well known as computer-aided nutrition design [13, 14].

Bioinformatics enterprises can produce amount of nutrition molecules in less time period with low risk. New nutrition molecules are usually only developed when the particular nutrition target for those nutrition molecules actions have been identified and studied. The role of bioinformatics in nutrition discovery process as shown in Figure 3.2 [15].

### 3.2.1  NUTRITION TARGET IDENTIFICATION

The most important emphasis of existing bioinformatics methodologies is the extrapolation and identification of naturally active candidates, and mining and storage of associated information. Nutrition are usually only developed when the particular nutrition target for those nutrition molecules actions have been identified and studied. The number of possible targets for nutrition discovery and development method is exponentially increases [16]. The nucleotide configurations of individual genes can be define and classify by mining and warehousing of the human genome sequence using of bioinformatics, which are answerable for the coding of target proteins and also helps to identifying new targets that gives more potential for new nutrition molecules. This is the place where the human genome evidence is predictable to play a master role. Nutrition scientist is offered with an unwonted extravagance of choice as additional genes are identified and the nutrition discovery cycle suits more data-intensive. Bioinformatics permits the identification and analysis of a number of biological nutrition targets; therefore predictable to significantly upsurge the breath of possible nutrition in the pipelines of healthy food and nutrition industries [17–19].

### 3.2.2   NUTRITION TARGET VALIDATION

Bioinformatics furnishes the plan and algorithm to predict new nutrition targets and to store and manage available nutrition target information. Subsequently, the discovery of possible nutrition targets, there is an irrelevant requirement to create a strong relationship between a supposed target and disease of interest. The initiation of such a key association provides a basis for the nutrition development process [20]. This procedure is recognized as target authentication and which is the province where bioinformatics is playing a significant role. Nutrition target authentication benefits to moderate the potential for disappointment in the clinical testing and endorsement phases [21].

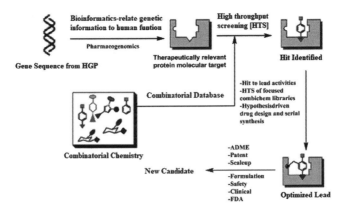

**FIGURE 3.2**   Role of bioinformatics in nutrition discovery process.

### 3.2.3 COST REDUCTION

The current high cost of nutrition discovery and development process is a major basis for concern among healthy food and nutrition industries. The healthy food and nutrition industries having aim to reduce the high failure rate in the nutrition discovery process with increasing productivity so that more number of nutrition molecules able to hit the market. The high price of numerous stages of clinical trials performances as preventive factors for amount of nutrition, which can be developed by healthy food and nutrition enterprises, and therefore choosing the compounds with the best probabilities for authorization is complicated. The overall costs of nutrition discovery and development process consist of total overheads from discovery to approval. In addition, some studies of nutrition discovery and development have been comprised of the costs of failed nutrition and the costs for commercialization. Nearby is also a cost accompanying with the extended process, start from discovery all the way to final approval. The nutrition discovery process accelerates through the advances in bioinformatics, starts with identification of nutrition target and validation by means of docking, assay development, and virtual-high-throughput screening (v-HTS). These all are the objectives of identifying new potential chemical entities. Bioinformatics helps to identify and analyzes one or more biological nutrition targets and thus expected to speeds the rate of growth of potential nutrition in the pipelines of healthy food and nutrition industries [22, 23].

### 3.2.4 NEW NUTRITION DEVELOPMENT

The major botheration healthy food and nutrition industry is that their some collateral costs, which include commercialization costs, litigation, nutrition-recall costs, and general costs to society. Commercialization costs for new potential nutrition are high mostly since recent "new" nutrition accepted are fundamentally functional duplications of nutrition that already exist and which are to be about $250 million per permitted nutrition [24]. Imitator nutrition are marketed to grip the illness to which nutrition are previously having; hence it is a requirement of the alliance that can fascinate the consideration of together physicians and patients who even now have use the similar medication. Bioinformatics is a proper route to provide new approaches and opportunities to the healthy food and nutrition industry for efficiently discovery of potential nutrition targets and develop innovative nutrition. If nutrition is not marketed in opposition with previously present counterparts, their marketed costs are predictable to fall significantly [25–27].

### 3.2.5   BARRIERS IN NUTRITION DESIGN PROCESS BY BIOINFORMATICS

Bioinformatics determined attempt will not make any impeccable changes in the nutrition discovery and development method because the practice of bioinformatics studies is relatively novel. On the foundation of the cost of nutrition till present, bioinformatics has not finished any significant influence for nutrition discovery. The healthy food and nutrition companies are constantly giving proof of increasing costs and removals of nutrition from the market after they had been approved and commercialized because of multiple documented cases of adverse nutrition reactions [28].

### 3.3   CHEMO-INFORMATICS

Chemo-informatics is the usage of computer and informational methods to intercept chemical problems, which accent the manipulation of chemical structural information. Since chemo-informatics is relatively new, there is no extensive diminish on accurate spelling. It is also known as cheminformatics, chemo-informatics, and chemical informatics. The term cheminformatics was first defined as "the involvement of evidence assets to alter data into evidence, and evidence into knowledge, for the intentional use of creation better choices quicker in the place of nutrition lead identification and optimization" by F. K. Brown [29]. Chemo-informatics is used in healthy food and nutrition industries in the process of nutrition discovery, such as virtual screening and quantitative structure-activity relationship (QSAR). An important representative of the methods in chemo-informatics is that it must be appropriate to a large amount of data and information (number of molecules). The data can only be processed and analyzed by computer methods, which also depend on computational power. The word "chemo-informatics" acquired shape only little years ago but quickly increased extensive usage. Here are a few of the first definitions of this term: "A critical part of the nutrition discovery process is the usage of information technology and management" [30, 31].

### 3.3.1  DIFFERENT ASPECTS OF CHEMO-INFORMATICS

Numerous of the chemo-informatics methodologies were introduced during the 1960s to 1970s and were execute into software systems that are here and now extensively applied and being unceasingly refined. Across the world, many research groups from the many days are still enthusiastically hunting further developments, and also many newer groups have participated in this area of research, elevating it with new ideas and novel software systems. The below sections show as an introduction to chemo-informatics and deliver a common impression on simple techniques and some precise fields from programming chemical compounds, loading, and searching data in databases, to constructing and examining data [32].

## 3.4  REPRESENTATION OF CHEMICAL COMPOUNDS AND REACTIONS

The input and output of chemical structures in computer curriculums and databases need precise approaches for their demonstration in computer-readable form. Structure illustrations also form the foundation for approaches used for (sub) structure examining in databases. Researchers may have miscellaneous ideas and insights when speaking about the structural evidence of chemical compounds. A compound may be categorized by its name, by its two-dimensional sketch, or by provided that a three-dimensional molecular model. This pyramid also reproduces the various levels of modification for on behalf of chemical structures in electronic forms preliminary from linear representations to two-dimensional chemical graphs and 3D structure demonstrations culminating in molecular planes. Linear representations show the structure of a chemical compound as a linked sequence of appeals and numbers. The greatest widespread linear notations are SLN (SYBYL line notation), SMILES (simplified molecular-input line-entry specification), ROSDAL (representation of organic structures description arranged linearly), and WLN (Wiswesser Line Notation). While initial linear representations were created before the entrance of the computer, it was rapidly understood that the density of such string representations is well suitable to account for the exclusive computational and storage assets [33].

Compared to SMILES, WLN, and ROSDAL is nearly related to the native conception of organic chemists and is grounded on a few important rules to change a structure into a character string. SMILES were generated in 1986 by David Weininger for chemical data treating and have established extensive delivery as a widespread chemical nomenclature for the demonstration and conversation of chemical structure [34].

## 3.5 APPLICATIONS OF BIOINFORMATICS

### 3.5.1 MOLECULAR MEDICINE

The human genome is having great possessions in the arenas of biomedical investigation and clinical medicine. Every ailment has a genetic component, and it may be inherited or because of the body shows the responses to an environmental stress which causes alterations in the genome. For example, diabetes, cancers, heart disease. The accomplishment of the human genome meaning that we can hunt for the genes straight related with diverse ailments and starts to recognize the molecular foundation of these ailments more evidently. This novel information of the molecular mechanisms of ailment will permit improved treatments, cures, and even prevention tests to be established [35].

### 3.5.2 PERSONALIZED MEDICINE

Clinical medicine will become more personalized with the development of the field of pharmacogenomics. This is the learning of how a person's genetic inheritance disturbs the body's comeback to nutrition. At present, some nutrition fails to develop it to the market since a small percentage of the clinical patient population display adverse effects to nutrition owing to sequence deviations in their DNA. As an end, possibly lifesaving nutrition never makes it to the marketplace [36].

### 3.5.3 PREVENTATIVE MEDICINE

Through the precise particulars of the genetic mechanisms of ailments being unraveled, the expansion of analytical tests to measure a person's vulnerability to diverse ailments may developed a discrete authenticity. Precautionary activities such as alteration of lifestyle or having treatment at the initial imaginable points when they are more likely to be fruitful could consequence in enormous developments in our scuffle to overcome disease [37].

### 3.5.4   GENE THERAPY

Gene therapy is the method used to treat, cure, or even escape disease by changing the appearance of a person's genes. Presently, this field is in its initial stage with clinical trials for numerous dissimilar types of cancer and other diseases ongoing [38].

### 3.5.5   NUTRITION DEVELOPMENT

At present, all nutrition on the market target only about 500 proteins. With an enhanced thoughtful of disease mechanisms and by means of computational tools to recognize and authenticate new nutrition targets, more precise medicines that act on the foundation, not simply the symptoms, of the ailment can be established. These extremely precise nutrition potential to have lesser side effects than many of today's medicines [39].

### 3.5.6   MICROBIAL GENOME APPLICATIONS

Microbes are permeating, that is they are inventing universally. They have been originating surviving and thriving in excesses of heat, cold, radiation, salt, acidity, and pressure. They are existing in the environment, our bodies, the air, food, and water. Conventionally, use has been prepared of a diversity of microbial belongings in the baking, preparing, and food industries. The appearance of the complete genome sequences and their potential to deliver a greater vision into the microbial world and its capabilities could have extensive and far attainment inferences for environment, health, energy, and industrial solicitations [40].

### 3.5.7   BIOTECHNOLOGY

Some archeon and the bacterium have possible for practical solicitations in industry and government-funded ecological remediation. These microbes increase in water temperatures overhead the boiling point and consequently, may deliver the DOE, the Department of Defense, and private industries with heat-stable enzymes appropriate for use in industrial developments. Other scientifically valuable microorganisms are of high manufacturing attention as a research object meanwhile it is useful by the chemical industries for the biotechnological manufacturing of the amino acid, lysine. The ingredient is used as a foundation of protein in animal nutrition. Lysine is one of the significant amino acids in animal nutrition. Biotechnologically manufactured lysine is complementary to nourish spirits as a basis of protein, and is a substitute to soybeans, meat, and bone meal [41].

### 3.5.7.1   ANTIBIOTIC RESISTANCE

Scientists have been examining the genome of a bacterium. They have exposed a virulence section prepared of a numeral of antibiotic-resistant genes that may donate to the bacterium's transformation from innocent gut bacteria to a threatening invader. The finding of the region, recognized as a pathogenicity island, might deliver valuable markers for detecting pathogenic strains and benefit to found controls to prevent the spread of infection inwards [42].

### 3.5.7.2   FORENSIC ANALYSIS OF MICROBES

Researchers used their genomic tools to help discriminate among the strains of a rod-shaped bacterium that was applied in the summer of 2001 terrorist attack in Florida with that of carefully associated anthrax strains [43].

### 3.5.7.3   THE AUTHENTICITY OF BIO-WEAPON CREATION

Researchers have lately manufactured the virus poliomyelitis by means of completely artificial means. They did this by means of genomic data obtainable on the Internet and materials from a mail-order chemical source. The investigation was sponsored by the US Department of Defense as part of a bio-warfare comeback program to suggestion to the world the authenticity of bio-weapons. The scientists also confidence for their work will dissatisfy officials from ever-restful courses of vaccination. This project has been happened with very mixed spirits [44].

### 3.5.7.4   EVOLUTIONARY STUDIES

The sequencing of genomes from all three fields of life, eukaryote, bacteria, and archaea by means of those evolutionary educations can be achieved in a pursuit to regulate the tree of life and the last worldwide common ancestor [45].

### 3.5.7.5   CROP IMPROVEMENT

Relative genetics of the plant genomes has revealed that the organization of their genes has continued more preserved over evolutionary time than was before supposed. These assumptions suggest that evidence acquired from the model crop schemes can be useful to recommend enhancements to other food crops. At contemporary the complete genomes of watercress and rice are obtainable [46].

### 3.5.7.6   INSECT RESISTANCE

Genes from *Bacillus thuringiensis* that can control a numerous of thoughtful pests have been efficaciously conveyed to cotton, maize, and potatoes. This novel capability of the plants to fight insect attack by means of that the quantity of pesticides being used can be reduced and later the nutritional excellence of the crops are augmented [47].

## 3.6   BIOINFORMATICS SOFTWARE AND DATABASES

A widespread composing of software, databases, and web services straight associated to nutrition discovery maintained by the Swiss Institute of Bioinformatics [48]. These are roughly grouped into:

1. Databases;
2. Chemical structure representations;
3. Molecular modeling and simulation;
4. Homology modeling to conclude the structure of a protein directed by a homolog of recognized structure;
5. Binding site prediction;

6.  Docking;
7.  Screening for nutrition candidates;
8.  Nutrition target prediction;
9.  Ligand design;
10. Binding free energy estimation;
11. QSAR;
12. ADME toxicity.

Numerous software packages are authoritative and free, and sustained by well-known organizations. These comprise databanks like ChEMBL and SwissSideChain, software tools such as UCSF Chimera, which is a visualization as well as platform for software developers fascinated in structural biology, SwissSimilarity for virtual screening, SwissBioisostere for ligand design, SwissTarget. Prediction, SwissSideChain to simplify experiments that enlarge the protein collection by presenting non-natural amino acids, and SwissDock for docking nutrition candidates (small molecules) on proteins. While some software are commercial, e.g., CHARMM, and PyMOL (Schrödinger), they characteristically have free versions for students and teachers [49, 50].

## 3.7  CONCLUSIONS

Most significant research fights in the arena of informatics encompass sequence alignment, protein structure alignment, gene finding, genome assembly, protein structure prediction, prediction of gene expression and protein-protein interactions, and the modeling of evolution. Bioinformatics allows the identification and analysis of more and more biological nutrition targets; thus expected to greatly increase the breadth of potential nutrition in the pipelines of healthy food and nutrition companies. Nutrition target validation benefits to modest the potential for failure in the clinical testing and approval phases. Prediction, SwissSideChain to facilitate experiments that expand the protein repertoire by introducing non-natural amino acids and SwissDock for docking nutrition candidates (small molecules) on proteins. Although some software are commercial, e.g., CHARMM, and PyMOL (Schrödinger), they typically have free versions for students and teachers.

## KEYWORDS

- **bioinformatics**
- **food and nutrition discovery**
- **new chemical entity**
- **nutrition design**
- **protein data bank**
- **quantitative structure-activity relationship**
- **representation of organic structures description arranged linearly**
- **simplified molecular input line entry specification**

## REFERENCES

1. Fenstermacher, D., (2005). Introduction to bioinformatics. *Journal of the American Society for Information Science and Technology, 56*(5), 440–446.
2. Akalın, P. K., (2006). Introduction to bioinformatics. *Molecular Nutrition and Food Research, 50*(7), 610–619.
3. Aoki-Kinoshita, K. F., (2008). An introduction to bioinformatics for glycomics research. *PLoS Computational Biology, 4*(5).
4. Krawetz, S. A., & Womble, D. D., (2003). *Introduction to Bioinformatics: A Theoretical and Practical Approach.* Springer Science & Business Media.
5. Pearson, W. R., (2013). An introduction to sequence similarity (homology) searching. *Current Protocols in Bioinformatics, 42*(1), 3.1.1–3.1.8.
6. Ewens, W. J., & Grant, G. R., (2006). *Statistical Methods in Bioinformatics: An Introduction.* Springer Science & Business Media.
7. Lesk, A., (2019). *Introduction to Bioinformatics.* Oxford University Press.
8. Gu, J., & Bourne, P. E., (2009). *Structural Bioinformatics* (p. 44). John Wiley & Sons.
9. Chou, K. C., (2015). Impacts of bioinformatics to medicinal chemistry. *Medicinal Chemistry, 11*(3), 218–234.
10. Dibyajyoti, S., Bin, E. T., & Swati, P., (2013). Bioinformatics: The effects on the cost of nutrition discovery. *Galle Medical Journal, 18*(1).
11. Qiao, L., & Zhang, Y., (2014). Application of CADD on multi-target nutrition R&D in natural products. *Z Hongguo Z Hongyao Magazine, z Hong Guo Chinese Medicine Magazine, China Journal of Chinese Materia Medica, 39*(11), 1951–1955.
12. Huang, H. J., et al., (2010). Current developments of computer-aided nutrition design. *Journal of the Taiwan Institute of Chemical Engineers, 41*(6), 623–635.
13. Xia, X., (2017). Bioinformatics and nutrition discovery. *Current Topics in Medicinal Chemistry, 17*(15), 1709–1726.
14. MacCuish, J. D., & MacCuish, N. E., (2010). *Clustering in Bioinformatics and Nutrition Discovery.* CRC Press.
15. Wishart, D. S., (2005). Bioinformatics in nutrition development and assessment. *Nutrition Metabolism Reviews, 37*(2), 279–310.

16. Chan, J. N., Nislow, C., & Emili, A., (2010). Recent advances and method development for nutrition target identification. *Trends in Pharmacological Sciences, 31*(2), 82–88.

17. Campillos, M., et al., (2008). Nutrition target identification using side-effect similarity. *Science, 321*(5886), 263–266.

18. Schenone, M., et al., (2013). Target identification and mechanism of action in chemical biology and nutrition discovery. *Nature Chemical Biology, 9*(4), 232.

19. Bantscheff, M., & Drewes, G., (2012). Chemoproteomic approaches to nutrition target identification and nutrition profiling. *Bioorganic and Medicinal Chemistry, 20*(6), 1973–1978.

20. Chen, X. P., & Du, G. H., (2007). Target validation: A door to nutrition discovery. *Nutrition Discov. Ther., 1*(1), 23–29.

21. Wise, A., Gearing, K., & Rees, S., (2002). Target validation of G-protein coupled receptors. *Nutrition Discovery Today, 7*(4), 235–246.

22. Veselovsky, A., & Ivanov, A., (2003). Strategy of computer-aided nutrition design. *Current Nutrition Targets-Infectious Disorders, 3*(1), 33–40.

23. Song, C. M., Lim, S. J., & Tong, J. C., (2009). Recent advances in computer-aided nutrition design. *Briefings in Bioinformatics, 10*(5), 579–591.

24. Kesselheim, A. S., Hwang, T. J., & Franklin, J. M., (2015). *Two Decades of New Nutrition Development for Central Nervous System Disorders*. Nature Publishing Group.

25. DiMasi, J. A., et al., (2010). Trends in risks associated with new nutrition development: Success rates for investigational nutrition. *Clinical Pharmacology and Therapeutics, 87*(3), 272–277.

26. DiMasi, J. A., (2001). Risks in new nutrition development: Approval success rates for investigational nutrition. *Clinical Pharmacology and Therapeutics, 69*(5), 297–307.

27. Mathieu, M. P., Evans, A. G., & Hurden, E. L., (1990). *New Nutrition Development: A Regulatory Overview*. PAREXEL International Corporation.

28. Katara, P., (2013). Role of bioinformatics and pharmacogenomics in nutrition discovery and development process. *Network Modeling Analysis in Health Informatics and Bioinformatics, 2*(4), 225–230.

29. Vandewiele, N. M., et al., (2012). Genesys: Kinetic model construction using chemo-informatics. *Chemical Engineering Journal, 207*, 526–538.

30. Andújar, I., et al., (2018). Use of chemo-informatics to identify molecular descriptors of auxins, cytokinins, and gibberellins. *J. Appl. Bioinforma. Comput. Biol., 7*(2), 2.

31. Ayyapan, G., (2013). Chemo Informatics QSAR analysis of nitroaromatic compounds toxicity. *Int. J. Innov. Res. Sci. En. Technol., 2*(2), 372–375.

32. Mishra, R., & Mishra, B., (2015). *Nutrition Discovery Tools: Computer-Aided Molecular Design and Chemo-Informatics, 4*(2), 261–272.

33. Gómez-Bombarelli, R., et al., (2018). Automatic chemical design using a data-driven continuous representation of molecules. *ACS Central Science, 4*(2), 268–276.

34. Reymond, J. L., & Awale, M., (2012). Exploring chemical space for nutrition discovery using the chemical universe database. *ACS Chemical Neuroscience, 3*(9), 649–657.

35. Taylor, R. C., (2010). An overview of the Hadoop/MapReduce/HBase framework and its current applications in bioinformatics. In: *BMC Bioinformatics*. Springer.

36. Liu, L., et al., (2016). An overview of topic modeling and its current applications in bioinformatics. *Springer Plus, 5*(1), 1608.

37. Maji, P., & Pal, S. K., (2012). *Rough-Fuzzy Pattern Recognition: Applications in Bioinformatics and Medical Imaging* (p. 3). John Wiley & Sons.

38. Wienbrandt, L., (2014). The FPGA-based high-performance computer RIVYERA for applications in bioinformatics. In: *Conference on Computability in Europe*. Springer.
39. DiMasi, J. A., & Grabowski, H. G., (2012). *R&D Costs and Returns to New Nutrition Development: A Review of the Evidence*. Oxford University Press Oxford.
40. Buermans, H., & Den, D. J., (2014). Next-generation sequencing technology: Advances and applications. *Biochimica et Biophysica Acta (BBA)-Molecular Basis of Disease, 1842*(10), 1932–1941.
41. Rittmann, B. E., & McCarty, P. L., (2012). *Environmental Biotechnology: Principles and Applications*. Tata McGraw-Hill Education.
42. Laxminarayan, R., et al., (2013). Antibiotic resistance-the need for global solutions. *The Lancet Infectious Diseases, 13*(12), 1057–1098.
43. Schmedes, S. E., Sajantila, A., & Budowle, B., (2016). Expansion of microbial forensics. *Journal of Clinical Microbiology, 54*(8), 1964–1974.
44. Bonomo, I., (2019). Ebola disease as a possible bio-weapon: A creativity or a prospect? *American Journal of Medical Science and Chemical Research, 2*(02), 01–03.
45. Laland, K., et al., (2014). Does evolutionary theory need a rethink? *Nature News, 514*(7521), 161.
46. Rafalski, J. A., (2010). Association genetics in crop improvement. *Current Opinion in Plant Biology, 13*(2), 174–180.
47. Onstad, D. W., (2013). *Insect Resistance Management: Biology, Economics, and Prediction*. Academic Press.
48. Goto, N., et al., (2010). BioRuby: Bioinformatics software for the Ruby programming language. *Bioinformatics, 26*(20), 2617–2619.
49. Pabinger, S., et al., (2014). MEMOSys 2.0: An update of the bioinformatics database for genome-scale models and genomic data. *Database*.
50. Dai, L., et al., (2012). Bioinformatics clouds for big data manipulation. *Biology Direct, 7*(1), 43.

# CHAPTER 4

# Functional Foods and Ingredients, Supplements, Nutraceuticals, and Superfoods Uses and Regulation

LUIS E. ESTRADA-GIL,[1] JUAN C. CONTRERAS-ESQUIVEL,[2]
MARCO A. MATA-GÓMEZ,[3] ADRIANA C. FLORES-GALLEGOS,[1]
ALEJANDRO ZUGASTI-CRUZ,[4] RAÚL RODRÍGUEZ-HERRERA,[5]
MAYELA GOVEA-SALAS,[6] and J. A. ASCACIO-VALDÉS[1]

[1]*Bioprocesses and Bioproducts Group, Food Research Department,
School of Chemistry, Autonomous University of Coahuila, Saltillo – 25280,
Coahuila, México, E-mail: alberto_ascaciovaldes@uadec.edu.mx
(J. A. Ascacio-Valdés)*

[2]*Laboratory of Glycobiotechnology, School of Chemistry,
Autonomous University of Coahuila, Saltillo – 25280, Coahuila, México*

[3]*School of Engineering and Science, Tecnológico de Monterrey,
Puebla – 5718, 72453, México*

[4]*Laboratory of Immunology, School of Chemistry,
Autonomous University of Coahuila, Saltillo – 25280, Coahuila, México*

[5]*Laboratory of Molecular Biology, Food Research Department,
School of Chemistry, Autonomous University of Coahuila, Saltillo – 25280,
Coahuila, México*

[6]*Laboratory of Nanobioscience, School of Chemistry,
Autonomous University of Coahuila, Saltillo – 25280, Coahuila, México*

## ABSTRACT

Nutraceuticals, functional foods, ingredients, and dietary supplements are novel products that are being analyzed and studied due to their health benefits and functional interactions in the human body as they contain bioactive compounds that have specific functionality in our systems, such components may be carotenoids, polyphenols, essential fatty acids (EFAs), pigments, dietary fibers and probiotics among other bioactive compounds. These biological compounds of interest contained within these products possess anti-inflammatory, antiallergic, antimicrobial, analgesic, and prebiotic properties that show to have positive effects in the prevention of many diseases such as cancer, cardiovascular diseases, metabolic syndrome, and diabetes type 2. Regulation of these products has also been present in many parts of the world, such as the United States, European Union (EU), and Japan, where they possess their classification and need to pass a vigorous and strict process for the product to have its health claim accepted.

## 4.1  INTRODUCTION

As time goes on and technology keeps on moving forward, innovation in all the fields of science keeps ongoing, and this is no exception for food science and technology, where new products are developed to satisfy new demands in our population. Not only this generation of consumers looks to nourish with the natural nutrients that our foods give us, but they also seek to get another benefit from what they are consuming; an example of this can be functional foods, nutraceuticals, supplements, and smart foods.

Agriculture producers, pharmaceutical companies, and food process companies have begun producing this kind of product that provides the benefit of preventing disease and helping them treating some other conditions that currently affect the population. These products are abundant in bioactive compounds that have a positive effect on the body or specific tissues or cells.

These bioactive compounds differ from nutrients in many ways as they are usually not present in large quantities in our alimentation, and also not taken into consideration for our nutritional guidelines [1].

These compounds can be pigments, antioxidants, carotenoids, and essential fatty acids (EFAs) that can be added to many food matrixes to make a functional food or ingredient. Antioxidant compounds are beneficial in this area due to their functional properties that help to prevent coronary diseases, as described in studies such as the French paradox, hypertension, and even mental disorders can be treated with antioxidant compounds making functional foods an excellent prevention treatment [2–5].

## 4.2   NOVEL HEALTH FOOD PRODUCTS

### 4.2.1   WHAT IS FUNCTIONAL FOOD?

Functional foods are any food or food ingredient that may provide a health benefit while also nourishing our body with the essential macronutrients. For an ingredient or food to be called functional food, we need to demonstrate that it can help prevent diseases or may as well help to treat them [6, 7]. A fundamental trait is that functional foods must remain as food, and their effects should be present by eating the usually consumed amount of the ingredient in a diet [8].

### 4.2.2   WHAT IS A NUTRACEUTICAL?

The word nutraceutical comes from a mix of the words "pharmaceutical" and "nutrition" it was first coined by Dr. Stephen DeFelice in 1989. Moreover, according to DeFelice, we can define a nutraceutical as a food or ingredient that can provide medicinal or health benefits, including the prevention or treatment of certain diseases. Usually, nutraceuticals come as pills, capsules, or microencapsulated products that can protect the main compound [9].

### 4.2.3   WHAT IS A NUTRITIONAL SUPPLEMENT?

A nutritional supplement is a product from which we can obtain micro or macronutrients to satisfy an augmented dietary need. Dietary supplements are not drugs used to cure diseases, so they cannot possess a claim of "reduces pain" or "prevents heart diseases" [10].

### 4.2.4   WHAT IS A SUPERFOOD?

By definition, a superfood needs to be of 100% organic origin and have a high amount of nutrients in just small portions. They are called this way for being an excellent source of antioxidants and essential nutrients. However, superfoods are just a marketing plan as most of these so-called superfoods claim to have health benefits, including the prevention of diseases. However, none of these benefits are scientifically proven [11].

## 4.3   HOW ARE THESE PRODUCTS USED TO PROMOTE HEALTH BENEFITS?

Food is a basic human necessity as everyone needs to satisfy this physiological need to get through the day and accomplish his or her daily goals. However, the population usually does not worry about what food they eat; provoking arises in coronary and metabolic diseases due to high calorie and low nutritional intake from fast food [12, 13].

Functional foods, nutraceuticals, and supplements are an alternative to the regular diet, and they provide us with additional health benefits. Studies show that consumption of these foods can prevent and treat some diseases; we are going to talk about this in-depth in the following sections.

Research has proved that there is a relationship between the functional components inside the food, health, and well-being. Some of the most common functional ingredients are dietary fibers, EFAs, polyphenols, and carotenoids [14].

### 4.3.1   DIETARY FIBER

Usually, dietary fiber is found in fruits, but mostly in the peel portion of them; vegetables and cereals are also an excellent source for dietary fiber. There are two kinds of fiber, soluble and insoluble; soluble fiber goes through a fermentation process inside the gastrointestinal tract (GIT), and this influences the metabolism of carbohydrates and fats. Insoluble fibers shorten the transit time inside the GIT, preventing constipation [15]. Dietary fiber is useful in the prevention of colon cancer, coronary heart disease, hypertension, obesity, and diabetes [16].

### 4.3.2   ESSENTIAL FATTY ACIDS (EFAS)

EFAs are a group of nutrients that constitute a crucial role in our system, providing structural sustain to the cell membrane and having essential roles in our health as nutraceuticals, with mostly cardioprotective functions [17, 18]. We can find EFA's in seeds, seafood (mostly salmon, tuna, and shellfish), and some fruits and vegetables (such as avocado).

### 4.3.3 POLYPHENOLS

Polyphenols or phenolic acids are one of the most abundant non-energetic compounds found inside plants. In our diet, their total dietary intake can be as high as 1g/d; this intake is around ten times greater than that of vitamin C and 100 times higher than that of vitamin E and carotenoids [19].

Research has proven that polyphenols can be used to treat and prevent certain diseases, giving them an essential role in pharmaceutical, food, and cosmetics industries; among those properties, we can find an anti-carcinogenic, antimicrobial, anti-inflammatory, cardioprotective, and antioxidant activity [20].

Polyphenols tend to reside in many food matrixes, such as fruits, vegetables, peels, seeds, leaves, and beverages like wine. Agricultural by-products such as the peel of the fruit can contain high amounts of polyphenol concentration, as shown in rambutan (*Nephelium lappaceum* L.) and pomegranate (*Punica granatum*) both show a significant number of polyphenols in their peels making them an asset [21, 22].

### 4.3.4 CAROTENOIDS

Carotenoids are a group of pigments and precursors of vitamin A found in various products, many of which are red, orange, and dark green colored vegetables and fruits. Humans or animals do not synthesize them, but they are present in the organism through food ingestion. However, their bioavailability, absorption (can vary from 5 to 50%), breakdown, transport, and storage dramatically depend on the food matrix from which we may obtain them as well as the association of carotenoids with other compounds [23]. Some carotenoids, such as lycopene or β-carotene, have a microcrystalline form, which makes them less available comparing them to those entirely immersed in lipid droplets [24]. Among the functional properties of carotenoids, we can find antioxidant, anti-carcinogenic, cardio-protector, and photo-protection [25].

These are just some of the essential functional ingredients that we can find among many others that constitute most of the functional foods, nutraceuticals, and supplements developed in recent times.

## 4.4  DISEASE PREVENTION AND TREATMENT

Several studies demonstrate that food intake is one of the main pillars in the prevention of chronic diseases, as they are of long duration and slow progression. Diabetes, heart attacks, coronary diseases, cancer, and respiratory illness are the leading causes of mortality in the world.

Nevertheless, as Hippocrates once said, "let food be the medicine and medicine be the food" food possesses antioxidants, carotenoids, pigments, fiber, and EFAs, among other components that can help us treating and preventing diseases. In the following section, we are going to discuss this even further.

### 4.4.1  DIABETES TREATMENT

Diabetes type 1 represents 5–10% of the cases of diabetes around the world, and it characterizes in the destruction of pancreatic β-cell that eventually will leave the body with no insulin and causing polyuria, polydipsia, polyphagia, weight loss, dehydration, loss of electrolytic balance and ketoacidosis.

Diabetes type 2 represents 90–95% of the cases, and it characterizes as being a combination of low amounts of pancreatic β-cell or impaired insulin signaling and insulin resistance. In most individuals, diabetes is present even before a diagnosis. In the long term, it can cause the risk of significant complications if not treated properly (necrosis, diabetic foot, and loss of sight, among others) [26].

In both types of diabetes, diet is the focus of the treatment, by regulating the amount of carbohydrates on our menu we can control the disease, recently functional foods and their bioactive ingredients have been considered as a way of managing and preventing diabetes. Focusing on diabetes type 2 polyphenols may be an appropriate, functional ingredient, nutraceutical, or supplement for the treatment of this disease [27].

Among the benefits of polyphenols in the treatment of diabetes, we can include:

- Inhibit α-glucosidase and α-amylase, therefore, lowering the digestion and intestinal absorption of carbohydrates by inhibiting the intestinal $Na^+$ dependent glucose transporters. Some examples of polyphenols with these actions are green tea catechins and epicatechins, chlorogenic, ferulic, caffeic, and tannic acids, quercetin, and naringenin [28, 29].

- Lowering gluconeogenesis and glucose output of the liver, rising glucogenesis and glycogen content in the liver and glycolysis for glucose oxidation, thus regulating the carbohydrate metabolism; some examples of these polyphenols are ferulic acid (FA), hesperidin, and naringin [30, 31].
- Rising insulin-dependent glucose via GLUT4, and by doing this, it improves the glucose uptake in the muscle cell and adipocytes, lowering the glucose level in blood [32].
- Protecting pancreatic β-cells from oxidative damage and decreasing cell apoptosis, regulating production and secretion of insulin, thus improving the function and insulin action [33].

With insulin resistance, there are other risks besides having high glucose levels in the blood. We can also find that patients with diabetes usually develop cardiovascular risk, and in time developing coronary artery disease being this among the significant causes of mortality in diabetes type 2.

Polyphenol rich food and supplementation can play a vital role in the prevention and treatment of cardiovascular diseases being the central fact that their antioxidant capacities allow them to regulate lipids metabolism [27].

### 4.4.2   CARDIOVASCULAR DISEASE TREATMENT

According to the World Health Organization (WHO), cardiovascular disease is one of the main problems around the world in prevalence and mortality as it is the primary cause of death and morbidity in the world. Governments around the world have been launching public health measures to prevent new cases, many of which are just like any other measure in the world such as lowering salt and sodium intake from canned or prepackaged foods, exercising once a day, reducing saturated fat intake and mostly education on how to live a healthy lifestyle.

Nevertheless, still, with all these measures taken around the world, the prevalence of cardiovascular disease is one of the top problems, and patients who already present this profile need to be treated, pharmacological treatment of the disease has dramatically improved. However, nutraceutical, functional food, and supplementation treatment have been considered among the main therapeutical approaches that should be taken. Some examples include polyphenols, dietary fiber, protein, and unsaturated fatty acids.

Hypercholesterolemia characterizes by high levels of cholesterol in the blood that are involved in the pathogenesis of the atherosclerotic changes in the vascular wall. Over the years, there have been studies on how nutraceuticals are used in the treatment and prevention of hypercholesterolemia, for example, zymosan, a water-soluble; wall yeast polysaccharide, which is composed of β-glucan and mannan, has previously been shown capable of showing hypolipidemic effects on a mouse model.

Hypocholesterolemic effects can be attributed to its ability to form a gel and bind with cholesterol, thus preventing absorption. Another nutraceutical found to have both hypoglycemic and hypolipidemic properties is glucomannan, which swells in the presence of water, and it forms a gel that delays gastric emptying time this way, decreasing postprandial glucose surge and insulin usage. A decrease of insulin also suppresses hepatic cholesterol synthesis through the reduction of the insulin-induced 3-hydroxy-3-methylglutaryl coenzyme A (HMG-CoA) reductase activity [34].

### 4.4.3 CANCER PREVENTION

Cancer is a significant health problem worldwide it is the second leading cause of death in the United States, the American Cancer Society calculates and estimates the numbers of new cancer cases and deaths that will occur in the United States every year, and for 2020 there are 1,806,590 new cases and 606,520 deaths projected [35].

According to the National Cancer Institute, "cancer is a name given to a collection of related diseases." All kinds of cancer began with cells dividing without stopping and spread into the surrounding tissues of the body. Cancer can start in almost any part of the body, as we made up of trillions of cells.

Usually, the typical cycle of a human cell is to grow and divide to form new cells in the body as they are needed. Once the cells grow old or become damaged, they go through a process called apoptosis, where old cells that are not required anymore die. However, when cancer develops, this process is interrupted, and cells that should go through an apoptosis process don't, and they start dividing and forming new cells that are not needed.

Treatment for cancer depends on the type of cancer presented and how advanced it may be. Some examples of cancer need a combination of medications or just one treatment such as chemotherapy, radiation therapy, immunotherapy, targeted therapy, or hormone therapy.

Many of these treatments tend to affect patients physically and psycho-logically due to the significant stress to which the patients are exposed, due to not only cancer but also the treatment can be substantial on their immune system. Therefore, prevention should be the primary approach that public health agencies and the population should practice, some bioactive ingre-dients and compounds in functional foods and nutraceuticals are proven to have a beneficial effect in the prevention of cancer.

### 4.4.3.1   LYCOPENE

Lycopene is a 40-carbon atom, open-chain hydrocarbon containing 11 conju-gated and 2 non-conjugated double bonds assigned in a linear array. Lyco-pene can be found in a wide variety of fruits such as watermelon, grapefruit, tomato, and apricots, among others, it is responsible for giving the fruits their characteristically deep red color. Amongst its biological activities they possess, we can find the antioxidant activity, cell-to-cell communication, and growth control, but it has no provitamin A action [36].

Lycopene is a more powerful antioxidant than α and β-carotene in preventing the cell growth of various cancer cell lines. In mice models, lycopene has shown anti-carcinogenic roles in mammary gland, liver, skin, and lungs [37].

### 4.4.3.2   PROBIOTICS AND PREBIOTICS

Probiotics are microorganisms that live in our GIT that have positive effects on health when consumed. They can be found in fermented products and dietary supplements; there are a wide variety of probiotics, but the most commonly known are *Lactobacillus* and *Bifidobacterium*. Meanwhile, prebiotics are non-digestible food components that these probiotic microor-ganisms use to stimulate their growth and activity.

Probiotics such as *Lactobacterium acidophilus* and *Bifidobacterium longum* have been studied and researched for their positive effects against cancer. Probiotics also inhibit putrefactive intestinal bacteria with enzymatic functions that generate carcinogenic substances from dietary components and change procarcinogens into carcinogens. Colorectal proliferation is a biomarker used to measure colon cancer risk; probiotics used in tests proved that their intervention tended to reduce colorectal proliferation in polyp patients [38].

*Bacillus polyfermenticus* is a kind of probiotic bacteria that can withstand digestive enzymes, gastric acid, and bile salts, and it is used to treat a variety of intestinal disorders due to its main attribute of producing the antimicrobial agent bacteriocin. Oral administration to patients shows how it stimulates IgG production and modulates the number of $CD4^+$ and $CD^8$, N.K. cells. The antiproliferative effects of *Bacillus polyfermenticus* are reported in Caco-2 colon cancer cells. The action mechanism by which B.P. inhibits the growth of cancer cells, including colon, breast, cervical, and lung cancers, is by reducing the expression levels of ErbB2 and ErbB3 protein and mRNA. Overexpression of ErbB2 and ErbB3 is observed in many human cancers [39].

Omega 3—they are a group of polyunsaturated fatty acids, also called linolenic acid. The most important and recurrent are α linolenic acid, eicosapentaenoic acid (EPA) and docosahexaenoic acid (DHA). Supplementation with omega-three fatty acids can be beneficial in reducing inflammation and cardiovascular diseases. People with diets high in omega-3 fatty acids could experience a lower prevalence of some types of cancer [40].

### 4.4.3.3   FLAVONOIDS

Flavonoids are part of the polyphenol family that is synthesized in plants. They have several beneficial roles in maintaining our health, including antiviral, antitoxic, antifungal, antibacterial, antiallergic, anti-inflammatory, and antioxidant activities [41]. Regarding anti-cancer activity, there have been studies that show a positive impact on the prevention and therapy of many types of cancer like ovarian, colon, lungs, prostate, pancreatic, and many more. Polyphenols such as epigallocatechin-3-gallate, which can be found in green tea, act as anti-cancer agents through the activation of transcription systems. Flavonoids' action mechanisms consist of the inhibition of proliferation, inflammation, invasion, metastasis, and activation of apoptosis [40].

Some examples of this are: luteolin and genistein, inhibit proliferation by inducing G1 and G2/M phase arrest in oral and pharyngeal cancer cells. Quercitin and flavan-3-ol induced apoptosis through activation of bad and hypophosphorylated retinoblastoma, thus preventing invasion and metastasis in pharyngeal carcinoma cells. Preclinical studies also show that quercetin protects against prostate cancer through inhibition of invasion, migration, and signaling through epidermal growth factor receptor, NFkB, and AP-1 inducing apoptosis [42].

## 4.5 REGULATIONS AROUND THE WORLD

### 4.5.1 REGULATIONS IN THE UNITED STATES

In the United States, the Food and Drug Administration (FDA) is responsible for protecting the health of the American population by ensuring the food and drug safety, efficacy, and security of human and veterinary products. It is responsible for regulating foods such as dietary supplements, bottled water, food additives, infant formulas, and other food products.

Focusing on how the FDA regulates products, such as dietary supplements, functional foods, and nutraceuticals, these last ones are taken into consideration as dietary supplements by the FDA. Current FDA regulations provide several provisions needed to communicate the health benefits the food products provide to the final costumers.

#### 4.5.1.1 NUTRIENT CONTENT CLAIM

Nutrient claims explicitly characterize the level of a nutrient in a food. These claims describe the amount of nutrients the product possesses by using words such as "free," "low," and "high," as well as "more," "reduced," and "light." These content claims are permitted for calories, fat, saturated fat, cholesterol, sodium, sugar, vitamins, minerals, fiber, and protein [43].

> **Structure and Function Claim:** These claims are statements that describe the effect of a dietary supplement or food ingredient that may have on the structure and function of the body. Note that these claims should not be confused with health claims, as they require prior approval by the FDA and maybe only for food products or dietary supplements that go through several tests that prove their health effects. FDA requires that these claims be truthful, non-misleading, and sustained by the appropriate scientific data of the food or dietary supplement to ensure the accuracy and truthfulness of the claim [44].

> **Health Claim:** These claims describe the relationship between foods or dietary supplements and disease in their labeling after such statements have been reviewed and approved by the FDA. These health claims are directed to all the population or some subgroups such as the elderly or infant population and are intended to assist the final consumer in maintaining healthy dietary practices [44].

## 4.5.2   REGULATIONS IN THE EUROPEAN UNION (EU)

The European Union (EU) bases its food regulation on different principles; some of them are:

1. The principle of liberalization of free trade: meaning the elimination of any form of trade restriction and the free circulation of food;
2. The principle of food safety governance;
3. The principle of comitology or committee procedure;
4. The principle of law harmonization;
5. The principle of an integrated approach;
6. The principle of food safety within a public health perspective;
7. The principle of consumer protection;
8. The principle of food traceability or accountability or transparency;
9. The principle of responsibility of food manufacturers;
10. The principle of science-based decision-making [45].

### 4.5.2.1   DIETARY SUPPLEMENT ACCORDING TO THE EU

Regulation (EC) No 1925/2006 of the European Parliament and of the council of 20 December 2006 dietary supplements are defined as concentrated sources of nutrients or other substances with a nutritional or physiological effect, intended to complement the regular diet, it could be used alone or with a combination. Dietary supplements should be marketed in dose form such as pills, tablets, capsules, or liquids in measured doses. The EU recognizes the following as dietary supplements: amino acids, enzymes, pre, and probiotics, EFAs, fiber, and various bioactive compounds.

### 4.5.2.2   NOVEL FOODS ACCORDING TO THE EU

Regulation (EC) No 258/97 of the European Parliament and of the council of 27 January 1997 Concerning novel foods and novel food ingredients defines them as the following novel foods are foods or food ingredients which have not been used to a significant degree for human consumption in the EU. This regulation applies to food or food ingredients from microorganisms, fungi, algae, plant, or animal ingredients.

### 4.5.2.3 NUTRACEUTICALS: ACCORDING TO THE EU

Regulation (EC) No 1924/2006 of the European Parliament and of the council of 20 December 2006, the EU has adopted the use of nutritional and health claims made on foods. Regulations apply to any product that claims to have an existing relationship between food or one of its constituent and health. Health claims need to comply with the following requirements: shall not be false, give rise to doubt about the safety or the nutritional adequacy of other foods, encourage the excess consumption of food, they shall not imply that a balanced and varied diet cannot provide adequate quantities of a nutrient and refer to changes in body functions that could give rise to fear in the consumer.

## 4.5.3 FUNCTIONAL FOOD REGULATION IN JAPAN

Japan is one of the oldest countries with a history of using the health benefits of the food as they are well known for their longevity over 65 years old. In 1991 the Ministry of health, Labor, and Welfare introduced the system called "foods for specified health uses" (FOSHU), and later in 2015, they introduced a new system called New Functional Foods; each one has its regulations and normative to take into consideration.

### 4.5.3.1 FOSHU SYSTEM

FOSHU products are approved by the Consumer Affairs Agency, regarding health claims, structural, and functional claims are not approved in the FOSHU system. Most Health claims available are related to the GI tract or metabolic syndrome. FOSHU system needs clinical evidence with a restricted protocol and needs registration in the University Hospital Medical Information Network (UMIN) [46].

### 4.5.3.2 NEW FUNCTIONAL FOODS SYSTEM

These products require no approval by any government agency as companies regulate them; they can be launched into the market after 60 days of acceptance. Structural, functional health claims are available in this system. New Functional Foods system also needs registration in the UMIN, and it has a more flexible protocol as previous clinical results published by other workgroups can be used if they show evidence of the efficacy of the active component in the food [46].

## 4.6 CONCLUSION

Novel health food products have been widely studied as they provide the population with a lot of health benefits within their grasp as they can be purchased in many stores and supermarkets. The only thing that is necessary for people to know the benefits of these products is for them to get the information they need. Further studies of agricultural by-products are also needed as they can also be used in the elaboration of nutraceutical and cosmetic products as they possess many of the properties previously mentioned that are not contained in the pulp of the fruit.

## KEYWORDS

- **docosahexaenoic acid**
- **eicosapentaenoic acid**
- **essential fatty acids**
- **European Union**
- **Food and Drug Administration**
- **gastrointestinal tract**
- **nutrition labeling and education act**
- **World Health Organization**

## REFERENCES

1. Siân, A., & Paul, F., (2016). Nutrition and health. In: *Reference Module in Food Sciences* (pp. 1–6). https://doi.org/10.1021/ie50204a022.
2. Cheng, H. S., Ton, S. H., & Abdul, K. K., (2017). *Ellagitannin Geraniin: A Review of the Natural Sources, Biosynthesis, Pharmacokinetics and Biological Effects* (Vol. 16). Springer Netherlands. https://doi.org/10.1007/s11101-016-9464-2.
3. Renaud, S., & De Lorgeril, M., (1992). Wine, alcohol, platelets, and the French paradox for coronary heart disease. *Lancet, 339*(8808), 1523–1526. https://doi.org/10.1016/0140-6736(92)91277-F.

4. Quiñones, M., Miguel, M., & Aleixandre, A., (2012) The polyphenols, naturally occurring compounds with beneficial effects on cardiovascular disease. *Nutr. Hosp. Nutr. Hosp., 2727*(1), 76–8976. https://doi.org/10.3305/nh.2012.27.1.5418.

5. Bravo, L., (2009). Polyphenols: Chemistry, dietary sources, metabolism, and nutritional significance. *Nutr. Rev., 56*(11), 317–333. https://doi.org/10.1111/j.1753-4887.1998. tb01670.x.

6. Milner, J. A., (2000). Functional foods: The US perspective. *Am. J. Clin. Nutr., 71*(6), 1654–1659. https://doi.org/10.1093/ajcn/71.6.1654s.

7. Annunziata, A., & Vecchio, R., (2011). Functional foods development in the European market: A consumer perspective. *J. Funct. Foods, 3*(3), 223–228. https://doi. org/10.1016/j.jff.2011.03.011.

8. Ghosh, N., Das, A., & Sen, C. K., (2019). Nutritional supplements and functional foods. In: *Nutraceutical and Functional Food Regulations in the United States and around the World* (pp. 13–35). Elsevier Inc. https://doi.org/10.1016/b978-0-12-816467-9.00002-2.

9. Kalra, E. K., (2003). Nutraceutical-definition and introduction. *AAPS J., 5*(3), 1–2. https://doi.org/10.1208/ps050325.

10. Food Facts from the U.S. Food and Drug Administration, (2017). *Food Facts About Dietetic Supplements.* http://www.fda.gov/educationresourcelibrary (accessed on 22 December 2020).

11. Van, D. D. J. J., Plat, J., & Mensink, R. P., (2018). Effects of superfoods on risk factors of metabolic syndrome: A systematic review of human intervention trials. *Food Funct., 9*(4), 1944–1966. https://doi.org/10.1039/c7fo01792h.

12. Zagorsky, J. L., & Smith, P. K., (2017). The association between socioeconomic status and adult fast-food consumption in the U.S. *Econ. Hum. Biol., 27*, 12–25. https://doi. org/10.1016/j.ehb.2017.04.004.

13. Keller, A., O'Reilly, E. J., Malik, V., Buring, J. E., Andersen, I., Steffen, L., Robien, K., et al., (2020). Substitution of sugar-sweetened beverages for other beverages and the risk of developing coronary heart disease: Results from the Harvard pooling project of diet and coronary disease. *Prev. Med. (Baltim)., 131*, 105970. https://doi.org/10.1016/j. ypmed.2019.105970.

14. Abuajah, C. I., Ogbonna, A. C., & Osuji, C. M., (2015). Functional components and medicinal properties of food: A review. *J. Food Sci. Technol., 52*(5), 2522–2529. https:// doi.org/10.1007/s13197-014-1396-5.

15. Wanlapa, S., Wachirasiri, K., Sithisam-Ang, D., & Suwannatup, T., (2015). Potential of selected tropical fruit peels as dietary fiber in functional foods. *Int. J. Food Prop., 18*(6), 1306–1316. https://doi.org/10.1080/10942912.2010.535187.

16. Chen, J., Zhao, Q., Wang, L., Zha, S., Zhang, L., & Zhao, B., (2015). Physicochemical and functional properties of dietary fiber from maca (*Lepidium meyenii* Walp.) liquor residue. *Carbohydr. Polym., 132*, 509–512. https://doi.org/10.1016/j.carbpol.2015.06.079.

17. Simopoulos, A. P., (2003). Essential fatty acids in health and chronic diseases. *Forum Nutr., 56*, 67–70. https://doi.org/10.1093/ajcn/70.3.560s.

18. Das, U., (2006). Essential fatty acids: A review. *Curr. Pharm. Biotechnol., 7*(6), 467–482. https://doi.org/10.2174/138920106779116856.

19. Scalbert, A., Johnson, I. T., & Saltmarsh, M., (2005). *Polyphenols: Antioxidants and Beyond, 81*, 215–217.

20. Kailasapathy, K., (2009). *Encapsulation Technologies for Functional Foods and Nutraceutical Product Development.* CAB Reviews: Perspectives in Agriculture,

Veterinary Science, Nutrition and Natural Resources. https://doi.org/10.1079/PAVSNNR20094033.

21. Hernández-Hernández, C., Aguilar, C. N., Rodríguez-Herrera, R., Flores-Gallegos, A. C., Morlett-Chávez, J., Govea-Salas, M., & Ascacio-Valdés, J. A., (2019). Rambutan (*Nephelium lappaceum* L.): Nutritional and functional properties. *Trends Food Sci. Technol., 85*, 201–210. https://doi.org/10.1016/j.tifs.2019.01.018.

22. Zam, W., Bashour, G., Abdelwahed, W., & Khayata, W., (2014). Alginate-pomegranate peels' polyphenols beads: Effects of formulation parameters on loading efficiency. *Brazilian J. Pharm. Sci., 50*(4), 741–748. https://doi.org/10.1590/S1984-82502014000400009.

23. Gallagher, M. L., (2013). Intake: Nutrients and metabolism. In: Breyer, P. L., Noland, D., & Johnson, R. K., (eds.), *Krause's Food and the Nutrition Care Process* (pp. 32–129). Elsevier Inc: Barcelona.

24. Fiedor, J., & Burda, K., (2014). Potential role of carotenoids as antioxidants in human health and disease. *Nutrients, 6*(2), 466–488. https://doi.org/10.3390/nu6020466.

25. Mayne, S. T., & Taylor, S., (1996). Beta-carotene, carotenoids, and disease prevention in humans. *Carotenoids Dis. Humans, 5*, 690–701.

26. Franz, M. J., (2013). Medical nutritional treatment in diabetes mellitus and hypoglycemia of nondiabetic origin. In: Breyer, P. L., Noland, D., & Johnson, R. K., (eds.), *Krause's Food and the Nutrition Care Process* (pp. 675–710). Elsevier Inc.: Barcelona.

27. Bahadoran, Z., Mirmiran, P., & Azizi, F., (2013). Dietary polyphenols as potential nutraceuticals in management of diabetes: A review. *Journal of Diabetes and Metabolic Disorders*, 1–9. https://doi.org/10.1186/2251-6581-12-43.

28. Johnston, K., Sharp, P., Clifford, M., & Morgan, L., (2005). Dietary polyphenols decrease glucose uptake by human intestinal caco-2 cells. *FEBS Lett., 579*(7), 1653–1657. https://doi.org/10.1016/j.febslet.2004.12.099.

29. Kobayashi, Y., Suzuki, M., Satsu, H., Arai, S., Hara, Y., Suzuki, K., Miyamoto, Y., & Shimizu, M., (2000). Green tea polyphenols inhibit the sodium-dependent glucose transporter of intestinal epithelial cells by a competitive mechanism. *J. Agric. Food Chem., 48*(11), 5618–5623. https://doi.org/10.1021/jf0006832.

30. Dao, T. M. A., Waget, A., Klopp, P., Serino, M., Vachoux, C., Pechere, L., Drucker, D. J., et al., (2011). Resveratrol increases glucose-induced GLP-1 secretion in mice: A mechanism which contributes to the glycemic control. *PLoS One, 6*(6). https://doi.org/10.1371/journal.pone.0020700.

31. Jung, U. J., Lee, M. K., Jeong, K. S., & Choi, M. S., (2004). The hypoglycemic effects of hesperidin and naringin are partly mediated by hepatic glucose-regulating enzymes in C57BL/KsJ-Db/Db mice. *J. Nutr., 134*(10), 2499–2503. https://doi.org/10.1093/jn/134.10.2499.

32. Kumar, R., Balaji, S., Uma, T. S., & Sehgal, P. K., (2009). Fruit extracts of *Momordica charantia* potentiate glucose uptake and up-regulate Glut-4, PPARγ and PI3K. *J. Ethnopharmacol., 126*(3), 533–537. https://doi.org/10.1016/j.jep.2009.08.048.

33. Fu, Z., & Liu, D., (2009). Long-term exposure to genistein improves insulin secretory function of pancreatic β-Cells. *Eur. J. Pharmacol., 616*(1–3), 321–327. https://doi.org/10.1016/j.ejphar.2009.06.005.

34. Johnston, T. P., Korolenko, T. A., Pirro, M., & Sahebkar, A., (2017). Preventing cardiovascular heart disease: Promising nutraceutical and non-nutraceutical treatments

for cholesterol management. *Pharmacol. Res., 120*, 219–225. https://doi.org/10.1016/j.phrs.2017.04.008.

35. Siegel, R. L., Miller, K. D., & Jemal, A., (2020). Cancer statistics, 2020. CA. *Cancer J. Clin., 70*(1), 7–30. https://doi.org/10.3322/caac.21590.

36. Stahl, W., & Sies, H., (1996). Perspectives in biochemistry and biophysics lycopene: A biologically important carotenoid for humans? *Arch. Biochem. Biophys., 336*(1), 1–9.

37. Levy, J., Bosin, E., Feldman, B., Giat, Y., Miinster, A., Danilenko, M., & Sharoni, Y., (1995). Lycopene is a more potent inhibitor of human cancer cell proliferation than either A-carotene or β-carotene. *Nutr. Cancer, 24*(3), 257–266. https://doi.org/10.1080/01635589509514415.

38. Rafter, J., Bennett, M., Caderni, G., Clune, Y., Hughes, R., Karlsson, P. C., Klinder, A., et al., (2007). Dietary synbiotics reduce cancer risk factors in polypectomized and colon cancer patients. *Am. J. Clin. Nutr., 85*(2), 488–496. https://doi.org/10.1093/ajcn/85.2.488.

39. Lee, K. H., Jun, K. D., Kim, W. S., & Paik, H. D., (2001). Partial characterization of polyfermenticin SCD, a newly identified bacteriocin of bacillus polyfermenticus. *Lett. Appl. Microbiol., 32*(3), 146–151. https://doi.org/10.1046/j.1472–765X.2001.00876.x.

40. Aghajanpour, M., Mohamad, R. N., Obeidavi, Z., Akbari, M., Ezati, P., & Kor, N. M., (2017). Functional foods and their role in cancer prevention and health promotion: A comprehensive review. *Am. J. Cancer Research, 7*(4), 740.

41. Sheng, H., So, C., & Ton, H., (2017). *Ellagitannin Geraniin: A Review of the Natural Sources, Biosynthesis, Pharmacokinetics and Biological Effects* (Vol. 16). Springer Netherlands. https://doi.org/10.1007/s11101-016-9464-2.

42. Romagnolo, D. F., & Selmin, O. I., (2012). Flavonoids and cancer prevention: A review of the evidence. *J. Nutr. Gerontol. Geriatr., 31*(3), 206–238. https://doi.org/10.1080/21551197.2012.702534.

43. FDA, (1994). *Guide to Nutrition Labeling and Education Act (NLEA) Requirements.* https://web.archive.org/web/20071014014547/https://www.fda.gov/ora/inspect_ref/igs/nleatxt.html (accessed on 22 December 2020).

44. FDA, (2009). *Guidance for Industry: Evidence-Based Review System for the Scientific Evaluation of Health Claims.* https://www.fda.gov/regulatory-information/search-fda-guidance-documents/guidance-industry-evidence-based-review-system-scientific-evaluation-health-claims (accessed on 22 December 2020). https://doi.org/10.1089/blr.2009.9977.

45. Bragazzi, N. L., Martini, M., Saporita, T. C., Nucci, D., Gianfredi, V., Maddalo, F., Di Capua, A., et al., (2016). *Nutraceutical and Functional Food Regulations in the European Union.* Elsevier Inc. https://doi.org/10.1016/B9780128027806.000171.

46. Iwatani, S., & Yamamoto, N., (2019). Functional food products in Japan: A review. *Food Sci. Hum. Wellness, 8*(2), 96–101. https://doi.org/10.1016/j.fshw.2019.03.011.

# PART 2

# Extraction of Essential Compounds

# CHAPTER 5

# Non-Starch Polysaccharides Extraction from Cereal by-Products Applying Pulsed Electric Fields (PEF) as Pretreatment: An Overview

BRENDA C. RAMIREZ-GONZALEZ,[1] MARIA I. CASTILLO-SANCHEZ,[1,2] RODRIGO MACIAS-GARBET,[1] ENA D. BOLAINA-LORENZO,[1] ANA MAYELA RAMOS-DE-LA-PEÑA,[3] MERAB M. RIOS-LICEA,[4] VIVIANA C. GUILLERMO-BALDERAS,[1] PAULINA V. PARGA-HERNANDEZ,[1] OSCAR F. VÁZQUEZ-VUELVAS,[5] JULIO C. MONTAÑEZ,[2] EMILIO MENDEZ-MERINO,[4] and JUAN CARLOS CONTRERAS-ESQUIVEL[1]

[1]*Laboratory of Applied Glycobiotechnology, Food Research Department, School of Chemistry, Autonomous University of Coahuila, Saltillo – 25250, Coahuila, Mexico, E-mail: carlos.contreras@uadec.edu.mx (J. C. Contreras-Esquivel)*

[2]*Department of Chemical Engineering, School of Chemistry, Autonomous University of Coahuila, Saltillo City – 25250, Coahuila, Mexico*

[3]*Tecnológico de Monterrey, School of Engineering and Sciences, Av. Eugenio Garza Sada – 2501 Sur, Monterrey – 64849, Nuevo Leon, Mexico, E-mail: ramos.amay@tec.mx*

[4]*Sigma Alimentos Co., Torre Sigma, San Pedro Garza Garcia – 66269, Nuevo Leon, Mexico*

[5]*Faculty of Chemical Sciences, Universidad de Colima, Coquimatlan – 28400, Colima, Mexico*

## ABSTRACT

Non-starch polysaccharides have been extracted from cereals using conventional and emerging technologies. Most of the conventional extraction methods involve heating. Pulsed electric fields (PEF) is a non-thermal technology that has been used in the food industry, and it is a promissory alternative to thermal polysaccharide extraction. The application of PEF for polysaccharides extraction has been scarcely reported. Existing reports show that PEF can improve the extraction of these compounds from cereals, and it can be considered as an assistive technology for extraction processes.

## 5.1  INTRODUCTION

Cereals belong to the Gramineae plant family and are considered a basic and important food in the human and animal diet. They are beneficial for health as they contribute to cancer, heart chronic diseases, and diabetes prevention [1]. A cereal grain is composed by 83% endosperm, 3% germ 3%, and 14% aleurone, and pericarp [2]. For instance, Figure 5.1 shows a corn grain structure.

Arabinoxylans (AXs) are non-starch polysaccharides, a type of hemicellulose that is contained in cereals. AXs are found in aleurone, endosperm, and pericarp layer regions [3]. These polysaccharides are a source of dietary fiber with prebiotic and immunomodulatory activity [4]. Moreover, their antioxidant activity is attributed to ferulic acid (FA) contained in the structure. AXs have been extracted from various cereals such as wheat, corn, barley, sorghum, rice, and oats [5] by means of chemical, physical, and enzymatic methods alone or combined. However, the use of high temperatures (160–200°C) for extractive purposes increased furfural production with increasing treatment time [11].

Extraction of AXs by pulsed electric fields (PEF) is an alternative technique to the aforementioned methods. The application of high voltage electric pulses for a few microseconds causes electroporation of the cell membrane, allowing compound diffusion from inside the cell to the surrounding medium [6–8]. In this way, the recovery yields of the extraction processes are improved.

PEF technology avoids waste generation (alkali/acid solutions), prevents FA degradation and low AXs branched degree [9, 10]. Due to its attributes, PEF has been used for plant, animal, and microbial cell membranes permeabilization. PEF technology has been applied to cereals for electroporation of ground corn to improve the use of raw fiber, increasing its digestibility in pigs feed [12].

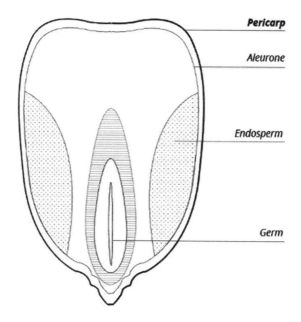

**FIGURE 5.1**    Structure of a corn grain.

The use of non-thermal PEF technology for the extraction of non-starch polysaccharides has been little studied. However, it is a promissory alternative for the extraction of AXs. The aim of this review is to present the use of PEF for the extraction non-starch polysaccharides such as AXs from several cereal grain sources.

## 5.2   ARABINOXYLANS (AXS)

Arabinoxylans are non-starch polysaccharides composed by a xylose main chain linked by β-1,4-glyosidic bonds. Arabinose is found on its side chains, linked in the position O-2 and O-3 [13] and glucuronic acid that are linked to xylose, as shown in Figure 5.2. AX can contain hydroxycinnamic acids such as FA bound to arabinose through an ester, bond on its part, ρ-cumaric acid can be found in lesser quantities [14]. FA can be found in diferulic and triferulic forms, linked through covalent bonds with other cell wall structures such as lignin and proteins [15, 16]. This is important in the cross-linking because they can potentially cross-link polysaccharide chains through a dimerization reaction [17].

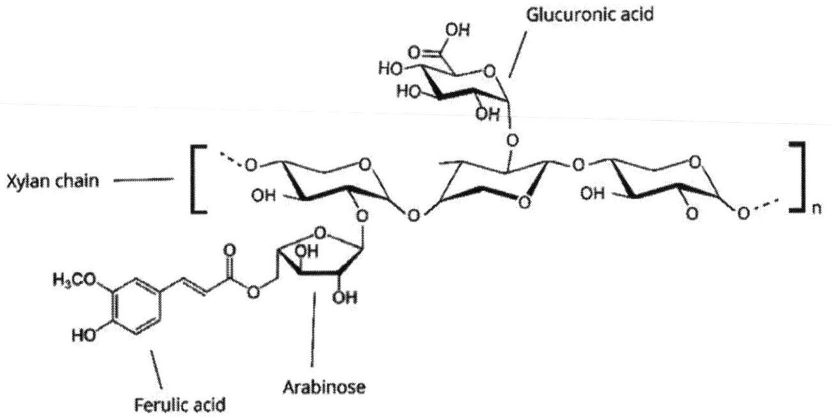

**FIGURE 5.2**    Structure of arabinoxylan.

AXs have been classified according to their solubility as water-extractable arabinoxylans (WEAXs) and water unextractable arabinoxylans (WUAXs). WEAXs are extractable at room temperature, and WUAXs require other treatments as they form covalent bonds with cell walls [18, 19]. These poly-saccharides have been extracted through chemical, physical, and enzymatic methods listed below:

1.  Extraction with chemical solvents. Since aqueous extraction of AXs is difficult due to high substitution degree, acids and alkalis are generally used [19–21].
2.  Extraction by enzymatic hydrolysis generally uses endo-xylanases [22].
3.  For mechanically assisted extraction, the application of compressed hot water, high-pressure steam, and the combination of some of them has also been tried. In the case of emerging technologies, microwave-assisted extraction and ultrasound-assisted extraction have been explored [16, 23].

## 5.3   PULSED ELECTRIC FIELD TECHNOLOGY

PEF is an emerging technology applied for the extraction of biopolymers and bioactive compounds from natural sources due to it is advantages in extraction processes [24, 25]. PEF is considered a non-thermal technology that takes advantage of the electroporation phenomenon by applying high voltage electric pulses for very shorts periods of time microsecond (µs). The permeabilization effect of the cell membrane leads to the extraction of compounds from biological matrices [6–8].

Duration and intensity of the electric field (E) applied determine the damage over the cell membrane. Effects can be reversible or irreversible permeabilization (lysis), or even lead to degradation of bioactive compounds, including those found in organelles [26, 27].

PEF have been useful for the recovery of polysaccharides from different sources such as cereals, fungi, algae, animals, fruits, vegetables, and there byproducts. However, it is not a fully exploited technology. According to Belwal et al. [25], PEF has low acceptability due to it has been tested with few solvents (compared with the Soxhlet method). Table 5.1 shows recent studies devoted to polysaccharides extraction from different sources.

**TABLE 5.1**  Application of Pulsed Electric Fields for the Extraction of Polysaccharides

| Source | Extract | Condition | Yield | References |
|---|---|---|---|---|
| Algae | Laminaria polysaccharide | Intensity 10 kV Pulse rate 50 Hz Solid liquid ratio 1:30 Pulse time 20 min | Compared to the traditional method by heating, it reduced extraction time by 50%, without heating. | [28] |
| *Morchella esculenta* | Intracellular polysaccharide | Intensity 18 kV/cm Pulse rate 7 Hz Liquid-solid ratio.27 ml / g | 56.03 µg/mL | [29] |
| Chicory roots | Inulin | Intensity 600 V/ cm Time treatment 10–50 µs Temperature 30–80°C | 11.65 g/100 mL Juice purity (87.1%) The same of conventional method, but the PEF treatment (10 ms) decreased the diffusion temperature by 10 to 15°C. | [30] |
| Yellow passion fruit peels | Pectin | 100 V Temperature 50°C Time 5 min pH 5 Peel: extractant (1:30) | Not improved, however time and temperature used were decreased. The characteristics of pectin are similar to those extracted by conventional method. | [31] |

**TABLE 5.1** *(Continued)*

| Source | Extract | Condition | Yield | References |
|---|---|---|---|---|
| *Rana temporaria chensinensis* | Rana temporaria polysaccharides | Intensity 20 kV/cm Pulse time 6 μs 0.5% KOH | 55.59% Higher than conventional method | [32] |
| Mulberries | Mulberries polysaccharides | Intensity 40 kV Temperature 80°C Pulse rate 10 Solid-liquid ratio 1:30 | 82% Conventional extraction only 67% | [33] |
| Mendong Fiber | Cellulose | Intensity 1.3 kV/cm Pulse rate 20 Hz Pulse time 30 second 50 g fiber/300 ml NaOH Solid-liquid ratio 60% | Better crystallinity is achieved than with alkaline extraction alone. | [34] |
| Apple pomace | Pectin | Intensity 15 kV/cm Pulse number 10 Relation solid-liquid 1:19 Temperature 62°C | 14.12% | [35] |
| Corn bran | Corn bran polysaccharides | Intensity 25 kV/cm Frequency 2080 Hz Liquid/solid ratio: 42 ml/g | PEF water-assisted extraction 6.4% PEF assisted enzyme extraction 15.36% | [36] |
| Corn silk | Corn silk polysaccharides | Intensity 30 kV/cm Liquid-solid ratio 50 ml/g Pulse duration 6 μs | 7.31 ± 0.15% | [37] |

## 5.4  APPLICATION OF PEF FOR EXTRACTION OF NON-STARCH POLYSACCHARIDES FROM CEREALS

### 5.4.1  *EXTRACTION OF CORN BRAN POLYSACCHARIDES BY PEF*

Pericarp or cornhusk is a byproduct of corn processing, such as starch and flour for tortilla industries, with a high content of AXs [14, 16]. Liu et al. [36] proposed the extraction of polysaccharides from pericarp using two methods. In the first case, the material was prepared on an aqueous medium for PEF treatment. In the second case, the PEF treated material was enzymatically processed afterwards using the enzymatic mix Celluclast® 1.5 L

(700 EGU/g). The highest performance conditions for Celluclast®-PEF treatment were: solid-liquid ratio 40:1 mL/g, electric field 25 kV/cm and frequency 2000 Hz. Results from Table 5.2 show that PEF and Celluclast® extraction had a higher yield compared to extraction using only PEF, that can be attributed to xylanolytic activity.

**TABLE 5.2**   Yield of Corn Bran Polysaccharides by PEF and Celluclast® [36]

| Tested Conditions | Fixed Conditions | Yield (%) | |
| --- | --- | --- | --- |
| | | **PEF** | **PEF/Celluclast** |
| Solid: liquid ratio (ml/g) 20, 30, 40, 50, 60 | 25 kV, 2000 Hz | 6 | 12 |
| Electric field (kV/cm) 10, 15, 20, 25, 30 | 40:1 ml/g, 2000 Hz | 8 | 14 |
| Frequency (Hz) 500, 1000, 1500, 2000, 2500 | 40:1 ml/g, 25 kV | 8 | 16 |

In the second stage, PEF conditions in an aqueous medium were optimized by applying a Box-Behnken design. According to the obtained response surface, the optimal parameters of the electric field, frequency, and solid-liquid ratio were 25 kV/cm, 2080 Hz, and 42:1 mL/g. These conditions, combined with Celluclast®, allowed the recovery of $15.36 \pm 0.25\%$ corn bran polysaccharide, 6.4% higher than the yield obtained using only PEF in an aqueous medium. The combination of PEF and enzymatic treatments gave the highest yield, as it improved polysaccharide extraction yield.

### 5.4.2   EXTRACTION OF BREWER SPENT GRAIN NON-STARCH POLYSACCHARIDES BY PEF

Brewer's spent grain (BSG) is known as a residue of the brewing industry, it is composed by hemicellulose, specifically AXs, lignin, beta-glucans, dietary fiber, proteins, and polyphenols. This material has been employed as raw material for the extraction of bioactive compounds using PEF technology [38].

Sajib [39] performed the non-starch polysaccharide extraction from BSG using PEF technology combined with autohydrolysis by autoclave, and PEF conditions for extraction were 20 kV/cm, 20 pulses, and 140 Hz. In addition, an experiment was carried out without PEF, using only autohydrolysis. The protocol for PEF application is shown in Figure 5.3. Polysaccharide yield using electric pulses was reported as 7.1%, while treatment without PEF obtained 6.68%. The combination of PEF and autohydrolysis showed a little increase in the yield of the non-starch polysaccharide compared to only autohydrolysis.

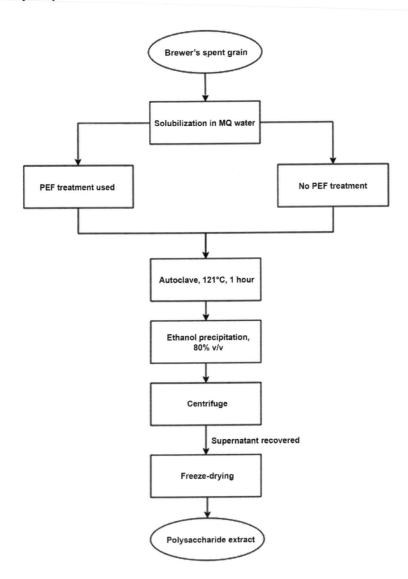

**FIGURE 5.3** Brewer's spent grain polysaccharide extraction process in the presence or absence of PEF treatment.

## 5.5 CONCLUSIONS

PEF technology has been used for polysaccharide extraction from several sources. However, its application for the extraction from cereals has been little studied. This technology has been used mainly as a pretreatment and combined with other treatments such as enzymatic and autohydrolysis promotes polysaccharide extraction of non-starch polysaccharides from cereal byproducts. This field of study has a high potential for the design of new efficient extractive processes in the field of cereal technology.

## KEYWORDS

- **arabinoxylan**
- **barley**
- **Celluclast®**
- **corn bran**
- **hemicellulose**
- **pulsating fields**

## REFERENCES

1. Masisi, K., Beta, T., & Moghadasian, M. H., (2016). Antioxidant properties of diverse cereal grains: A review on *in vitro* and *in vivo* studies. *Food Chem., 196*, 90–97.
2. Labuschagne, M. T., (2010). A review of cereal grain proteomics and its potential for sorghum improvement. *J. Cereal Sci., 84*, 151–158.
3. Carvajal-Millan, E., Rascon-Chu, A., Marquez-Escalante, J., Micard, V., Ponce, D. L. N., & Gardea, A., (2007). Maize extraction gum: Characterization and functional properties. *Carbohydr. Polym., 69*, 280–285.
4. Mendis, M., & Simsek, S., (2014). Arabinoxylans and human health. *Food Hydrocoll., 42*, 239–243.
5. Fadel, A., Mahmoud, A. M., Ashworth, J. J., Li, W., Ng, Y. L., & Plunkett, A., (2018). Health-related effects and improving extractability of cereal arabinoxylans. *Int. J. Biol. Macromol., 109*, 819–831.
6. Raso, J., Frey, W., Ferrari, G., Pataro, G., Knorr, D., Teissie, J., & Miklavcic, D., (2016). Recommendations guidelines on the key information to be reported in studies of application of PEF technology in food and biotechnological processes. *Innov. Food Sci. Emerg. Technol., 37,* 312–321.
7. Moussa-Ayoub, T. E., Jaeger, H., Youssef, K., Knorr, D., El-Samahy, S., Kroh, L. W., & Rohn, S., (2016). Technological characteristics and selected bioactive compounds of

*Opuntia dillenii* cactus fruit juice following the impact of pulsed electric field (PEF) pretreatment. *Food Chem., 210,* 249–261.

8. Mannozzi, C., Fauster, T., Hass, U., Romani, S., Rosa, D. M., & Jaeger, H., (2018). Role of thermal and electric field effects during the pretreatment of fruit and vegetable mash by pulsed electric fields (PEF) and ohmic heating (OH). *Innov. Food Sci. Emerg. Technol., 48,* 131–137.

9. Zhang, Z., Smith, C., & Li, W., (2014). Extraction and modification technology of arabinoxylans from cereal byproducts: A critical review. *Food Res. Inter., 65,* 423–436.

10. Sanchez-Bastardo, N., Romero, A., & Alonso, E., (2017). Extraction of arabinoxylans from wheat bran using hydrothermal processes assisted by heterogeneous catalysts. *Carbohydr. Polym., 160,* 143–152.

11. Rose, D. J., & Inglett, G. E., (2010). Production of feruloylated arabinoxylo-oligosaccharides from maize (*Zea mays*) bran by microwave-assisted autohydrolysis. *Food Chem., 119,* 1613–1618.

12. Web page Big Dutchman, (2020). https://www.bigdutchman.es/es/portal-es/ (accessed on 22 December 2020).

13. Ayala-Soto, F. E., Campanella, O. H., Serna-Saldivar, S. O., & Welti-Chanes, J., (2016). Changes in the structure and gelling properties of maize fiber arabinoxylans after their pilot scale extraction and spray-drying. *J. Cereal Sci., 70,* 275–281.

14. Lapierr, C., Pollet, B., Ralet, M. C., & Saulnier, L., (2001). The phenolic fraction of maize bran: Evidence for lignin-heteroxylan association. *Phytochemistry, 57,* 765–772.

15. Saulnier, L., Vigouroux, J., & Thibault, J. F., (1995). Isolation and partial characterization of feruloylated oligosaccharides from maize bran. *Carbohydr. Res., 272,* 241–253.

16. Yoshida, T., Tsubaki, S., Teramoto, Y., & Azuma, J. I., (2010). Optimization of microwave-assisted extraction of carbohydrates from industrial waste of corn starch production using response surface methodology. *Biores. Technol., 101,* 7820–7826.

17. Saulnier, L., Marot, C., Chanliaud, E., & Thibault, J. F., (1995). Cell wall polysaccharide interactions in maize bran. *Carbohydr. Polym., 26,* 279–287.

18. Morales-Ortega, A., Niño-Medina, G., Carvajal-Millan, E., Gardea-Béjar, A., Torres-Chavez, P., Lopez-Franco, Y., & Lizardi-Mendoza, J., (2013). Ferulated arabinoxylans from cereals. A review of their physico-chemical characteristics and gelling capability. *Rev. Fitotec. Mex., 36,* 439–446.

19. Kale, M. S., Hamaker, B. R., & Campanella, O. H., (2013). Alkaline extraction conditions determine gelling properties of corn bran arabinoxylans. *Food Hydrocoll., 31,* 121–126.

20. Anderson, C., & Simsek, S., (2018). Mechanical profiles and topographical properties of films made from alkaline extracted arabinoxylans from wheat bran, maize bran, or dried distillers grain. *Food Hydrocoll., 86,* 78–86.

21. Vidal, Jr. B. C., Rausch, K. D., Tumbleson, M. E., & Singh, V., (2009). Determining corn germ and pericarp residual starch by acid hydrolysis. *Cereal Chem., 86,* 133–135.

22. Beaugrand, J., Chambat, G., Wong, V. W., Goubet, F., Remond, C., Paes, G., Benamrouche, S., et al., (2004). Impact and efficiency of GH10 and GH11 thermostable endoxylanases on wheat bran and alkali-extractable arabinoxylans *Carbohydr. Res., 339,* 2529–2540,

23. Hromádková, Z., Kováčiková, J., & Ebringerová, A., (1999). Study of the classical and ultrasound-assisted extraction of the corn cob xylan. *Ind. Crops Prod., 9,* 101–109.

24. Okolie, L. C., Akanbi, T. O., Mason, B., Udenigwe, C. C., & Aryee, A. N. A., (2019). Influence of conventional and recent extraction technologies on physicochemical

properties of bioactive macromolecules from natural sources: A review. *Food Res. Inter., 116*, 827–839.

25. Belwal, T., Ezzat, M. S., Rastrelli, L., Bhatt, D. I., Daglia, M., Baldi, A., Devkota, P. H., et al., (2018). A critical analysis of extraction techniques used for botanicals: Trends, priorities, industrial uses and optimization strategies. *TrAC Trends in Anal. Chem., 100*, 82–102.

26. Toepfl, S., Heinz, V., & Knorr, D., (2005). Overview of pulsed electric field processing for foods. In: Sun, D. W., (ed.), *Emerging Technologies for Food Processing* (pp. 67–97). Elsevier, Oxford, UK.

27. Toepfl, S., (2011). Pulsed electric field food treatment- scale up from lab to industrial scale. *Procedia Food Sci., 1*, 776–779.

28. Liu, H., Feng, L. Q., & Liu, X. J., (2016). Comparative studies on extraction of laminarin by high voltage pulsed electric fields and traditional method. *Modern Chemical Industry, 7*, 75–78.

29. Liu, C., Sun, Y., Mao, Q., Guo, X., Li, P., Liu, Y., & Xu, N., (2016). Characteristics and antitumor activity of *Morchella esculenta* polysaccharide extracted by pulsed electric field. *Inter. J. Mol. Sci., 17*, 986.

30. Zhu, Z., Bals, O., Grimi, N., & Vorobiev, E., (2012). Pilot scale inulin extraction from chicory roots assisted by pulsed electric fields. *Inter. J. Food Sci. Technol., 47*, 1361–1368.

31. Oliveira, D. C. F., Giordani, D., Gurak, P. D., Cladera-Oliveira, F., & Marczak, L. D. F., (2015). Extraction of pectin from passion fruit peel using moderate electric field and conventional heating extraction methods. *Innov. Food Sci. Emerg. Technol., 29*, 201–208.

32. Yongguang, Y., Yuzhu, H., & Yong, H., (2006). Pulsed electric field extraction of polysaccharide from *Rana temporaria chensinensis* David. *Inter. J. Pharm., 312*, 33–36.

33. Jin, S. L., (2017). Optimization of extraction of polysaccharides from mulberries by using high-intensity pulsed electric fields. In: Kao, J. C. M., & Sung, W. P., (eds.), *Civil, Architecture and Environmental Engineering* (Vol. 1, pp. 741–745). CRC Press: Leiden, The Netherlands.

34. Suryanto, H., Fikri, A. A., Permanasari, A. A., Yanuhar, U., & Sukardi, S., (2018). Pulsed electric field assisted extraction of cellulose from mendong fiber (*Fimbristylis globulosa*) and its characterization. *J. Nat. Fibers., 15*, 406–415.

35. Yin, Y. G., Fan, X. D., Liu, F. X., Yu, Q. Y., & He, G. D., (2009). Fast extraction of pectin from apple pomace by high intensity pulsed electric field. *J. Jilin Univ. (Eng. Technol. Ed.), 39*, 1224–1228.

36. Liu, X. Y., Guo, X., & Lin, S. Y., (2016). Improvement of extraction rate of corn bran polysaccharide by high-voltage pulsed electric field. *J. Food Safety Qual., 7*, 2419–2425.

37. Zhao, W., Yu, Z., Liu, J., Yu, Y., Yin, Y., Lin, S., & Chen, F., (2011). Optimized extraction of polysaccharides from corn silk by pulsed electric field and response surface quadratic design. *J. Sci. Food Agric., 91*, 2201–2209.

38. Kumari, B., Tiwari, B. K., Walsh, D., Griffin, T. P., Islam, N., Lyng, J. G., Brunton, N. P., & Rai, D. K., (2009). Impact of pulsed electric field pretreatment on nutritional and polyphenolic contents and bioactivities of light and dark brewer's spent grains. *Inn. Food Sci. Emer. Technol., 54*, 200–210.

39. Sajib, M., (2017). *Preparation and Evaluation of Arabinoxylan Based Prebiotics*. MSC. Thesis, Division of Biotechnology, LTH, Lund University, Lund, Sweden.

# CHAPTER 6

# Research Progress on Application of Celluclast® as a Processing Aid for Pectin Extraction from Kiwifruit Pomace: A Mini Review

GABRIELA P. CID-IBARRA,[1] MARÍA G. RODRÍGUEZ-DELGADO,[1] EVA L. FERNÁNDEZ-RODRÍGUEZ,[1] CARLOS N. CANO-GONZALEZ,[1] ADRIANA J. SAÑUDO-BARAJAS,[2] JULIO C. MONTAÑEZ,[1] ANA MAYELA RAMOS-DE-LA-PEÑA,[3] ROSABEL VELEZ-DE-LA-ROCHA,[2] MERAB MAGALY RIOS-LICEA,[4] OSCAR F. VAZQUEZ-VUELVAS,[5] EMILIO MENDEZ-MERINO,[4] and JUAN CARLOS CONTRERAS-ESQUIVEL[1]

[1]School of Chemistry, Autonomous University of Coahuila, Saltillo – 25280, Coahuila, Mexico, E-mail: carlos.contreras@uadec.edu.mx (J. C. Contreras-Esquivel)

[2]Research Center for Food and Development, AC, Culiacan – 80110, Sinaloa, Mexico

[3]Tecnologico de Monterrey, School of Engineering and Science. Av. Eugenio Garza Sada – 2501 Sur, Monterrey, Nuevo Leon – 64849, Mexico

[4]Sigma Alimentos Co., Torre Sigma, San Pedro Garza Garcia – 66269, Nuevo Leon, Mexico

[5]Faculty of Chemical Sciences, Universidad de Colima, Coquimatlan – 28400, Colima, Mexico

## ABSTRACT

Celluclast® is an industrial enzyme preparation that is commonly used in the food and biorefinery industries. Recently, Celluclast® has played an important role in the extraction of pectin from various raw materials. This review is focused on the analysis of research works using Celluclast® to extract pectic substances from kiwifruit pomace. To date, a total of four papers were found with reference to the use of Celluclast® for pectin extraction. Kiwifruit pectin yield obtained with Celluclast® was lower than those found for other raw materials such as apple pomace or artichoke pomaces. Works on kiwifruit pectins are limited and more research on this topic needs to be carried out.

## 6.1  INTRODUCTION

Kiwi is a plant from the Yangtze valley of China. The variety *Actinidia deliciosa* is the most common, and its main characteristic is to have brown hairy skin and a bittersweet taste. The variety *Actinidia chinensis* has a hairless skin and a more yellowish color with a sweet and tropical flavor. Due to the agricultural progress of the XIX century, kiwifruit gained more attention due to it became economically profitable [1]. In 1904, the seed was taken to New Zealand, where the production is 80% is green kiwi and 20% gold varieties [2]. Kiwifruit can be consumed as raw fruit or processed to obtain food products or additives [3]. For instance, jams, preserves, fruit juice, wine, nectar, pulp products (fruit toppings), fruit leathers, dried fruit, candy (popular as canned fruit in China) can be obtained from kiwifruit [4]. In the processing to obtain various products from kiwi, most of the fruit is discarded, generating by-products. Kiwifruit waste comes in two forms. The first is the whole fruit that does not meet the quality standards for direct sale, and the second is kiwifruit pomace, which is the press cake residue after kiwifruit juice manufacture. The use of fruit wastes has been a great concern of fruit growers and manufacturers for at least 100 years [5]. Some of the uses of the pomace and by-products of the kiwifruit are citric acid production via fermentation, recovery of enzymes and cell wall components, source of vitamin C, animal feed, and pectin recovery [3, 6–9].

Among the by-products mentioned, pectin production stands out, which is a structural polysaccharide found in the cell walls of growing plants [10]. This pectic polymer is widely used in different areas such as the food industry for its gelling and stabilizing properties, and in the pharmaceutical industry because it favorably influences cholesterol levels in blood and acts as a natural prophylactic substance against poison [11]. In the food industry, pectin is used to jellify products, also it is used in confectionery fillings [12].

Pectin is composed of a linear chain of $\alpha$-1,4-D-galacturonic acid, this linear region is known as smooth region and a region known as pilose, with two subdivisions: rhamnogalacturonan I and II. The first one is mainly composed of homogalacturonan and the hairy region, the rhamnogalacturonan I has a backbone of α-(1,2)-linked L-rhamnosyl and α-(1,4)-linked D-galacturonosyl acid residues, with neutral sugars such as arabinose and galactose as the side chains.

Rhamnogalacturonan II shares with rhamnogalacturonan-I α-(1,2)-linked L-rhamnosyl and α-(1,4)-linked D-galacturonosyl acid and different sugars such as apiose, rhamnose, glucose, galactose, arabinose, fucose, xylose among others [13, 14].

Due to the applications of pectin, various methods to extract the polysaccharide have been explored, for instance, conventional techniques using mineral acids and innovative techniques (ultrasound-, microwave-, and enzyme-assisted extraction) [15]. Method selection for pectin extraction depends on several factors, including access to raw materials, water, energy, and effluent disposal at reasonable prices [16]. Disadvantages of pectin extraction by chemical methods include environmental pollution corrosion of reactors and equipment [17]. Enzymatic extraction seems to be more advantageous in terms of selectivity, energy consumption, and waste management [18].

In the fruit and vegetable processing industry, the use of enzymes is very popular for several purposes, such as shell softening (hydrolysis processes) and the release of polysaccharides from the plant cell wall. Another of the advantages of using macerating enzymes is the recovery of pulp, having a better juice yield, as in the case of kiwifruit [19].

Enzymes have been used for a long time to obtain pectic substances [20, 21]. It is common to use enzymatic preparations instead of purified enzymes, this due to the high costs that pure enzymes can generate. Among these enzymes is pectinesterase, which de-esterifies the methoxyl groups present in pectin. Endo-polygalacturonase hydrolyzes the region of low methoxylated homogalacturonan, whilst endo-pectin-lyase that hydrolyzes pectins with a high degree of methylation. Other enzymes that act on pectin are rhamnogalacturoanases, acetyl methyl esterases, α-L-arabinofuranosidases, among others [13]. It has been observed that enzymatic commercial preparations containing cellulases, in addition to other enzymes (polygalacturonases, xyloglucanases), may extract higher content of pectic substances than other treatments [2, 27]. This is due to pectic substances are released by the action of different kinds of enzymes by hydrolysis [22].

The aim of this mini-review is the analysis of research papers related to the use of Celluclast® as a processing aid for pectin extraction from kiwifruit pomace.

## 6.2 CELLUCLAST® AS PROCESSING AID FOR PECTIN EXTRACTION

Celluclast® is an enzymatic preparation manufactured by Novozymes (Denmark) and produced by a selected strain of filamentous fungi *Trichoderma reesei* [23]. This enzymatic preparation is used for cleaving down cellulose with cellulose-degrading enzymes. However, it has also been used successfully in the area of pectin technology to extract pectic substances from several raw materials (Table 6.1). The enzyme preparation Celluclast® 1.5 L has been used to extract pectin from several sources such as passion fruit, rapeseed cake, sugar beet pulp, apple pomace, lemon pomace and gold kiwifruit to extract pectin with yields in the range of 4 to 34% (dry basis). To date, only 13 research works have been carried out on the bioprocess of pectin extraction with Celluclast®.

**TABLE 6.1**  Sources and Pectin Yields by Using Celluclast®

| Sources of Pectin | Scientific Name | Yield (%) | References | Number of Articles |
|---|---|---|---|---|
| Sugar beet pulp | *Beta vulgaris* | 19.00 | [24] | 1 |
| Rapeseed cake | *Brassica napus L* | 6.85 | [25] | 1 |
| Passion fruit | *Passiflora edulis F. flavicarpa* | 7.16 | [26] | 1 |
| Chicory | *Cichorium intybus* | 18.95 | [27] | 1 |
| Apple pomace | *Malus domestica* | 18.95 | [28] | 4 |
|  |  | 15.30 | [29] |  |
|  |  | 21.24 | [30] |  |
|  |  | 22.71 | [31] |  |
| Artichoke | *Cynara scolymus* | 34.60 | [32] | 1 |
| Golden kiwifruit | *Actinidia chinensis* | 4.39 | [33] | 4 |
|  |  | 8.08 | [34] |  |
|  |  | 4.39 | [10] |  |
|  |  | 4.50 | [2] |  |

## 6.3   PECTIN EXTRACTION OF KIWIFRUIT POMACE WITH CELLUCLAST®

Table 6.2 shows the pectin yields obtained from kiwifruit pomace using Celluclast®, reported by Yulliarti et al. [34], which varies in a range of 6–8% (dry basis). It is relevant to observe that the obtained yield using 1.05 mL/kg of Celluclast® had a variation in the publications of the author, despite temperature and time were factors that remained constant. Of the four-kiwifruit articles, three of them had a performance of around 4%, while the article [34], showed a yield of 8%. Table 6.2 shows the pectin yield increased as more enzyme amount was added; however, in the third row, at higher enzyme concentration, lower pectin yields were obtained [34]. There are variations in the yield of kiwifruit pectin using the same conditions of both temperature and time. The amount of Celluclast® suitable for extraction is 1.05 (mL/kg) since higher or lower values showed lower yields. To date, there are a few studies of pectin extraction from kiwifruit. Jona et al. [35] performed a chemical and enzymatic extraction from kiwifruit and obtained a yield of 40% of pectic substances. Lodge et al. [36] obtained a yield of 0.85%, and Kawabata et al. [37] pointed out a kiwifruit pectin yield of 0.50–0.99%. Therefore, the pectin yields obtained from kiwifruit have been different in the analyzes performed [35–37].

**TABLE 6.2**   Effect of Celluclast® 1.5 L Content in the Pectin Yield from Kiwifruit Pomace

| Celluclast® (mL/kg) | Yield (%) | References |
|---|---|---|
| 0.10 | 6.58 | |
| 1.05 | 8.08 | [33] |
| 2.00 | 7.01 | |
| 1.05 | 2.14 | |
| 1.05 | 4.50 | [12] |

## 6.4   EXTRACTION OF KIWIFRUIT PECTIN WITH DIFFERENT COMMERCIAL ENZYME PREPARATIONS

Table 6.3 shows the pectin yields from kiwifruit using four different enzymatic preparations. In 2012, Yuliarti et al. [33] used Celluclast® 1.5 L, Cellulyve TR 400™, Cytolase CL™, and NS 33048™ to compare the results by using each enzymatic preparation alone and combined. The highest yield was obtained using the combination of Celluclast® 1.5 L, Cytolase CL™, and NS 33048™, yielding 5.34% and 5.40%. The results showed that the combination of these enzymatic preparations gave higher yields. However, these results were low compared to those obtained by Yuliarti et al. in 2011 [34], which was 8.08% using only Celluclast™. Conditions such as temperature (25°C) and time (30 min) remained constant [33].

**TABLE 6.3**   Effect of Enzyme Preparations on Pectin Yield from Kiwifruit Pomace

| Enzyme Extract | Enzyme Concentration (mL/kg) | Yield (%) |
|---|---|---|
| Celluclast® 1.5 L | 1.05 | 4.39 |
| NS 33048™ | 0.003 | NR* |
| Cellulyve TR 400™ | 0.15 | 4.82 |
| Cytolase CL™ | 0.06 | 4.00 |

*NR: No reported.

## 6.5   EXTRACTION METHODS USED TO OBTAIN PECTIN FROM KIWIFRUIT POMACE: ACID, HYDROTHERMAL, AND ENZYMATIC

Table 6.4 shows three extraction methods reported by Yuliarti et al. in 2015 [10]. When the enzymatic preparation Celluclast® 1.5 L was used, a yield of 4.48% was obtained, while the acid and the hydrothermal methods showed a yield of 3.83% and 3.62%, respectively. According to these results, the enzymatic method is the most effective to extract pectin from kiwifruit pomace, even more, effective than the acid method, where the temperature was higher than in the other two methods. Although the yield obtained with Celluclast® was slightly higher compared to the other two methods, the amount of pectin obtained was minimal. These values could be attributed to the type of pectin that is extracted from this source [2].

**TABLE 6.4**   Extraction Methods for Pectin Recovery from Kiwifruit Pomace: Acid, Hydrothermal, and Enzymatic (Celluclast® 1.5L).

| Extraction Method | Yield (%) | Time (min) | Temperature (°C) |
|---|---|---|---|
| Acid | 3.83 | 60 | 50 |
| Water | 3.62 | 30 | 25 |
| Enzymatic | 4.48 | 30 | 25 |

## 6.6   CONCLUDING REMARKS

The use of the Celluclast® enzyme preparation to obtain pectin from kiwi-fruit did not show a significant difference compared to the acidic and hydro-thermal methods used, showing only a slight increase in pectin yield. In this way, it can be concluded that, if the yields do not show a difference, the use of enzymes to obtain the polymer could be an option to use eco-friendly processes and thus contribute to reducing the waste generated. The use of enzymes to obtain pectin from kiwifruit is poorly studied, and it would be convenient to expand the experimentation to make comparisons, due to large differences in performance were detected, making comparisons difficult. The question if the use of enzyme preparations is good and determining the necessary amounts of enzyme to obtain the expected results and conduct experiments to study the ratio of the cost and the yield of product obtained. To date, by-products such as lemon pomace and apple pomace seem to be more effective for pectin extraction by enzymatic treatments with acceptable yields compared to those obtained from kiwifruit.

## KEYWORDS

- **D-galacturonosyl acid**
- **kiwifruit**
- **L-rhamnosyl**
- **pectic substances**
- **polygalacturonase**
- **pomace**
- ***Trichoderma ressei***

## REFERENCES

1. Richardson, D. P., Ansell, J., & Drummond, L. N., (2018). The nutritional and health attributes of kiwifruit: A review. *Eur. J. Nutr., 57*, 2659–2676.
2. Yuliarti, O., Goh, K. K. T., Matia-Merino, L., Mawson, J., Williams, M., & Brennan, C., (2015). extraction and characterization of pomace pectin from gold kiwifruit (*Actinidia chinensis*). *Food Chem., 187*, 290–296.
3. Kennedy, M. J., (1994). Apple pomace and kiwifruit: Processing options. *Australasian Biotechnol., 4*, 43–49.

4. Huang, H., & Ferguson, A. R., (2001). Review: Kiwifruit in China. *New Zeal. J. Crop. Hort., 29*, 1–14.

5. Kennedy, M., List, D., Lu, Y., Foo, L. Y., Robertson, A., Newman, R. H., & Fenton, G., Kiwifruit waste and novel products made from kiwifruit waste: Uses, composition and analysis. In: Linskens, H. F., & Jackson, J. F., (eds.), *Analysis of Plant Waste Materials. Modern Methods of Plant Analysis* (Vol. 20, pp. 121–152). Springer: Berlin.

6. Hang, Y. D., Luh, B. S., & Woodams, E. E., (1987). Microbial production of citric acid by solid-state fermentation of kiwifruit peel. *J. Food Sci., 52*, 226–227.

7. Baker, E. N., Boland, M. J., Calder, P. C., & Hardman, M. J., (1980). The specificity of actinidin and its relationship to the structure of the enzyme. *Biochim Biophys. Acta (BBA)-Enzymology, 616*, 30–34.

8. Dawson, D. M., & Melton, L. D., (1991). Two pectic polysaccharides from kiwifruit cell walls. *Carbohydr. Polym., 15*, 1–11.

9. Cano, M. P., & Marin, M. A., (1992). Pigment composition and color of frozen and canned kiwi fruit slices. *J. Agric. Food Chem., 40*, 2141–2146.

10. Yuliarti, O., Goh, K. K. T., Matia-Merino, L., Mawson, J., Williams, M., & Brennan, C., (2015). Characterization of gold kiwifruit pectin from fruit of different maturities and extraction methods. *Food Chem., 166*, 479–485.

11. Sriamornsak, P., (2003). Chemistry of pectin and its pharmaceutical uses: A review. *Silpakorn Univ. Int. J., 3*, 206–228.

12. Sakai, T., Sakamoto, T., Hallaret, J., & Vandamme, E. J., (1993). Pectin, pectinase, and protopectinase: Production, properties, and applications. *Adv. Appl. Microbiol., 39*, 213–294.

13. Contreras-Esquivel, J. C., (2003). *Purification and Characterization of Polygalacturonases from Aspergillus Kawachii*. PhD Dissertation, Universidad Nacional de La Plata. La Plata, Buenos Aires, Argentina.

14. Munarin, F., Tanzi, M. C., & Petrini, P., (2012). Advances in biomedical applications of pectin gels. *Int. J. Biol. Macromol., 51*, 681–689.

15. Marić, M., Grassino, A. N., Zhu, Z., Barba, F. J., Brnčić, M., & Rimac-Brnčić, S., (2018). An overview of the traditional and innovative approaches for pectin extraction from plant food wastes and by-products: Ultrasound-, microwaves-, and enzyme-assisted extraction. *Trends Food Sci. Technol., 76*, 28–37.

16. May, C. D., (1990). Industrial pectins: Sources, production and applications. *Carbohydr. Polym., 12*, 79–99.

17. Proverbio, E., & Bonaccorsi, L. M., (2002). Erosion-corrosion of a stainless steel distillation column in food industry. *Eng. Fail Anal., 9*, 613–620.

18. Dominiak, M., Søndergaard, K. M., Wichmann, J., Vidal-Melgosa, S., Willats, W. G. T., Meyer, A. S., & Mikkelsen, J. D., (2000). Application of enzymes for efficient extraction, modification, and development of functional properties of lime pectin. *Food Hydrocoll., 40*, 273–282.

19. Sharma, H. P., Patel, H., & Sugandha, (2017). Enzymatic added extraction and clarification of fruit juices: A review. *Crit. Rev. Food Sci. Nutr., 57*, 1215–1227.

20. Hiroshi, S., Takashi, A., Michiko, U., Kazutosi, N., & Akio, K., (1967). Nature of the macerating enzymes from *Rhizopus sp. J. Ferm. Technol., 45*, 73–85.

21. Kabli, A., (2007). Purification and characterization of protopectinase produced by *Kluyverumyces marxianus. J. King Abdulaziz Univ. Sci., 19*, 139–153.

22. Sakamoto, T., Hours, R., & Sakai, T., (1994). Purification, characterization, and production of two pectic transeliminases with protopectinase activity from *Bacillus subtilis*. *Biosci. Biotechnol. Biochem., 58*, 353–358.

23. Hjortkjaer, R. K., Bile-Hansen, B., Hazelden, K. P., McConville, M., McGregor, D. B., Cuthbert, J. A., Greenough, R. J., et al., (1986). Safety evaluation of celluclast, an acid cellulase derived from *Trichoderma reesei*. *Food Chem. Toxicol., 24*, 55–63.

24. Pacheco, M. T., Villamiel, M., Moreno, R., & Moreno, F. J., (2019). Structural and rheological properties of pectin's extracted from industrial sugar beet by-products. *Molecules, 24*, 392–408.

25. Jeong, H. S., Kim, H. Y., Ahn, S. H., Oh, S. H., Yang, I., & Choi, I. G., (2014). Optimization of enzymatic hydrolysis conditions for extraction of pectin from rapeseed cake (*Brassica napus* L.) using commercial enzymes. *Food Chem., 157*, 332–338.

26. Liew, S. Q., Chin, N. L., Yusof, Y. A., & Sowndhararajan, K., (2015). Comparison of acidic and enzymatic pectin extraction from *passion fruit* peels and its gel properties. *J. Food Process Eng., 39*, 501–511.

27. Panouillé, M., Thibault, J. T., & Bonnin, E., (2006). Cellulase and protease preparations can extract pectins from various plant byproducts. *J. Agric. Food Chem., 54*, 8926–8935.

28. Wikiera, A., Mika, M., & Grabacka, M., (2015). Multicatalytic enzyme preparations as an effective alternative to acid in pectin extraction. *Food Hydrocoll., 44*, 156–161.

29. Wikiera, A., Mika, M., Starzynska-Janiszewska, A., & Stodolak, B., (2015). Application of celluclast® 1.5L in apple pectin extraction. *Carbohydr. Polym., 134*, 251–257.

30. Dranca, F., Vargas, M., & Oroian, M., (2020). Physicochemical properties of pectin from *Malus domestica* 'fălticeni' apple pomace as affected by non-conventional extraction techniques. *Food Hydrocoll., 100*, 105383.

31. Dranca, F., Vargas, M., & Oroian, M., (2019). Optimization of pectin enzymatic extraction from *Malus domestica* 'fălticeni' apple pomace with celluclast 1.5L. *Molecules, 24*, 2158.

32. Sabater, C., Corzo, N., Olano, A., & Montilla, A., (2018). Enzymatic extraction of pectin from artichoke (*Cynara scolymus* L.) by-products using celluclast 1.5L. *Carbohydr. Polym., 190*, 43–49.

33. Yuliarti, O., Matia-Merino, L., Goh, K. T., Mawson, J., & Brennan, C., (2012). Characterization of gold kiwifruit pectin isolated by enzymatic treatment. *Int. J. Food. Sci. Technol., 47*, 663–649.

34. Yuliarti, O., Matia-Merino, Goh, K. K. T., Mawson, J. A., & Brennan, C. S., (2011). Effect of celluclast 1.5 L on the physicochemical characterization of gold kiwifruit pectin. *Int. J. Mol. Sci., 12*, 6407–6417.

35. Jona, R., & Fronda, A., (1997). Comparative histochemical analysis of cell wall polysaccharides by enzymatic and chemical extractions of two fruits. *Biotech. Histochem., 72*, 22–28.

36. Lodge, N., Nguyen, T. T., & Mcintyre, D., (1987). Characterization of a crude kiwifruit pectin extract. *J. Food Sci., 52*, 1095–1096.

37. Kawabata, A., Sawayama, S., & Uryu, K., (1974). A study on the contents of pectic substances in fruits, vegetable fruits, and nuts. *Jpn. J. Nutr. Diet., 32*, 9–18.

# CHAPTER 7

# Lectins from *Viscum album* (Mistletoe) Plants as Bioactive Compounds for Cancer Treatment

J. DANIEL GARCÍA-GARCÍA,[1] MELANY GARCÍA-MORENO,[1]
ROBERTO ARREDONDO-VALDÉS,[1] RODOLFO RAMOS-GONZÁLEZ,[1]
MÓNICA LIZETH CHÁVEZ-GONZÁLEZ,[1] JOSÉ LUIS MARTÍNEZ-
HERNÁNDEZ,[1] ELDA PATRICIA SEGURA-CENICEROS,[1]
MAYELA GOVEA-SALAS,[1] SANDRA CECILIA ESPARZA-GONZÁLEZ,[2] and
ANNA ILYINA[1]

[1]*Nanobioscience Group, Chemistry School,*
*Autonomous University of Coahuila, Saltillo, Mexico,*
*Tel.: +52 (844) 139-6175, E-mail: annailina@uadec.edu.mx (A. Ilyina)*

[2]*Odontology Faculty, Autonomous University of Coahuila, Saltillo, Mexico*

## ABSTRACT

Currently, lectins have acquired a great interest on the part of investigators due to the properties that they possess. Lectins are proteins present in plants, animals, and microorganisms. The main characteristic of this kind of protein is its ability to join sugar in a specific way. This ability made it a great tool in the treatment and diagnostics of diseases like cancer, and it has been used in another illness like diabetes or pains. Mistletoe lectins (MLs) can be considered as powerful treatments of different disorders and diseases. It has been reported the negative effects of lectin for the digestive system; however, the wide properties of lectins in cancer diseases can be a potential alternative for these diseases, which can be used as a dietary supplement. Due to mistletoe grows in pines and trees, it is considered as a hemiparasitic plant and it causes losses of money. However, mistletoe acquires nutrients like vitamins and minerals from its host, which made it rich in compounds of industrial interest. Although MLs from Europe are the most studied, American Mistletoe is an area for researches due to the low papers about lectin from this material and the greatest application of lectins. MLs represent an alternative option for using in different diseases. The purpose of this chapter is to analyze the potential of MLs as bioactive compounds and their current application.

## 7.1    INTRODUCTION

It is reported that plants do not poses immunologic system. Secondary metabolism is the responsible for defending plants against threats. One example of this defense are lectins, their main role in plants is to act as toxins or defense proteins that protect the host from phytopathogenic microorganisms, insects, and nematodes, and other predators such as plant-eating animals and humans. Lectins are macromolecules of a protein nature recognized for their ability to bind with carbohydrates of other molecules, reversibly, and specifically with specific sugar residues. This ability allows a molecular and cellular recognition of target objects that are involved in functions on biological systems; therefore, they appear in important biological settings and can be extremely useful analytical tools in organism-wide studies. The lectins are part of many biological organisms and, therefore, there are found in a variety of foods like grains, legumes, vegetables, and dairy products. The effects of lectin consumption can be very diverse, for example, some legume lectins are considered as a toxin or anti-nutritional as lentils, peas, fava beans, soybean, chickpea; nevertheless, these compounds are not inactivated by cooking or digestive processes, although there are process like fermentation which can reduce lectin content. The variety of lectins structures, recognition specificity of sugar moieties, and, therefore, activity, and application, is reported by several authors [1]. Europe has developed a wide investigation about lectin, and even lectins are used as an adjuvant in cancer therapy. It is believed Ricin is the first lectins discovered in the seeds of legumes, *Ricinus communis*, in 1888; however, there are reports that animal lectins were discovered before plant lectins, although many were not recognized as carbohydrate-binding proteins for many years. The present chapter aims to describe the basic properties of lectins and the methods of their extraction, with a special focus to the Mistletoe lectins (MLs) that has great potential for application in the regulation of different biological systems. Then, the properties and biochemical-biotechnological applications of MLs are analyzed.

## 7.2    LECTINS GENERALITIES

Lectins are carbohydrate-binding proteins, they are not part of humoral immune response and not possess catalytic activity, and therefore, they are considered neither antibodies nor enzyme. Glycoproteins are proteins that are modified after translation by carbohydrates addition [2]. Lectins can recognize sites of glycan residues in target systems and join reversibly to them [3, 4].

Lectins from animals were discovered earlier than from plants, although they were not recognized as proteins bound to carbohydrates. Kilpatrick [5] published a general overview of animal lectins, which discuss a historical overview of animal overview. In this chapter, he mentioned that Charcot and Robin described for the first time in 1853, a protein with the capability to agglutinate rattlesnake venom, similar observations were made by Leyden in 1872. These proteins were associated with eosinophil-mediated inflammation and were called Charcot-Leyden crystals; however, the first lectin from animal source was reported in 1902 by Flexner and Noguchi, and the pure form was published until 1980 from snake venom. Nowadays, it is believed that those proteins known as "Charcot-Leyden crystals" could be what today we call lectins. Considering this, and the fact that plant lectins were stated until 1888 as proteins from *Ricinus communis* seeds [5], the question "Was the first lectin from animal?" still unclear. Animal lectins are present in fishes [6–8], mussels [9], crustaceans [10], and snakes [11]. There are reports in which lectin is involved as part of the first defense against parasites in a water environment that promotes the parasitic diseases in fishes [12].

Unlike animals or humans, plants do not possess an immunological system, and also they cannot avoid or escape from animal predators; consequently, plants synthesize molecules as a mechanism for defending. Lectins participate in the defense against phytophagous insects and herbivorous animals, and probably against microorganisms as well. The role of lectin in plants relies on their carbohydrate-binding activity, which is presented in the cell surface of the microorganism and insects. Besides, they can have structural functions or cellular recognition on plasma membranes surface. Considering the great variety of carbohydrates associated with insects, bacteria, and viruses, there is no doubt that lectins in plants can be more than just one type in each plant; therefore, this is the interesting potential of lectin. As it has been mentioned, the properties of plant lectins are exogenous, and no specific endogenous role is reported clearly. This is the reason why lectins represent an excellent research area and the explanation of why plant lectins have different effects as inflammatory modulation [13], anticoagulant activity [14], antitumoral activity [15], among others. Currently, lectins from plants have been studied to identify the capability of differentiation malignant tumors according to the degree of glycosylation of cells [15]. *Viscum album* lectins have effects on cancer cells, which can be used as an alternative for disease treatment [16]. In fact, in Europe, MLs are used as an adjuvant in conjunction with chemotherapy and radiotherapy [10, 11]. Moreover, plant lectins have been evaluated in different cell lines *in vitro,* obtaining favorable results [17], which will be discussed later on.

Lectins can be differentiated according to the type of bond through which glycosylation is carried out: N-glycosidic and O-glycosidic bond [18]. The lectins are specific to carbohydrates, i.e., the molecular structure of target cells, and to the amino acid sequence. These can be found mostly in plants; some examples are shown in Table 7.1 [19]. However, they also can be found in animals [4–6] and microorganisms. The source of lectins is also the basis for their classification.

**TABLE 7.1**  Lectin from Different Plants and Animals

| Origen | Lectin Code | Source Latin Name | References |
|--------|-------------|-------------------|------------|
| Vegetal | ArtinM | *Artocarpus heterophyllus* | [20] |
| | BBL, BmoroL, BFL, BUL, BvvL | *Genero Bauhinia* | [21] |
| | PHA-E2L2, PHA-L4, PHA-E3L. | *Phaseolus vulgaris* | [22] |
| | LCaL | *Lonchocarpus campestris* | [23] |
| | Con A | *Canavalia ensiformis* | |
| | PCL | *P. cyrtonema* | [24] |
| | RIPsII | *Mistletoe Viscum album* | |
| | DVL | *Dioclea violacea* | [25] |
| | AIL | *Artocarpus integrifolia* | [26] |
| Animal | OnBML | *Oreochromis niloticus* | [26] |
| | Tn-pufflectin | *Takifugu niphobles* | [3] |
| | *Cs*LTL-1 | *Channa striatus* | [7] |
| | *Sm*LTL | *Scophthalmus Maximus* | [12] |
| | *Cs*BML1, *Cs*BML2, *Cs*BML3 | *Cynoglossus semilaevis* | [27] |
| | BLL | *Bothrops leucurus* | [11] |

Different lectins from various plants are known. In recent years, lectins have been isolated from a wide variety of plants, which, according to the maturity of the used plants, can be classified as merolectins, hololectins, chimericines, and superlectins. In addition, there is another classification, which is according to biological effect. For example, vegetable lectins have been divided as lectins of legumes, type II lectins associated with ribosomes, and lectins related to GNAt [24].

## 7.3   MISTLETOE PLANTS AS LECTIN SOURCE

### 7.3.1   GENERAL DESCRIPTION OF MISTLETOE

The plants which are known as "mistletoe," are native to Europe and western and southern Asia, also to America. It is believed that these plants came from Great Britain. The word mistletoe has an uncertain etymology. However, it is thought that it could be related to the German words "zein twig" with an uncertainly meaning. Mistletoe plants have been considered as "mythical plants," a symbol of male fertility [27]. Nowadays, the evergreen mistletoe is a symbol of fertility and good luck. There is a tradition of kissing under mistletoe branches that means a promise of marriage and a prediction of happiness [28]. Since old times, this plant has been used as a medicinal plant [29] as a treatment of menstruation pains, epilepsy, infertility, tumors, liver, and heart diseases, etc., as examples of traditional application of these plant [28, 30, 31].

The term "Mistletoe" is generally used to refers to plants with hemi-parasitic characteristics, such as growth on some trees, and with taxonomical similarity from five main families (*Misodendraceae, Viscaseae, Eremolepi-daceae, Amphorogynaceae,* and *Loranthaceae*) [32, 33]. Mistletoe families include about 100 species that could be found in Africa, Madagascar, Asia, and America [34]. The plants of these families are often considered as parasites because they kill trees during growth over them. Parasitic plants are characterized by the ability to grow and feed directly on other plants, which can grow either the roots or shoots of their hosts [31]. Hemiparasitic plants develop a specific strategy to acquire their resources combining parasitism and their own photosynthetic activity [30]. Vidal-Rusell and Nickrent [35] reported that families *Misodendraceae, Viscaseae, Eremolepidaceae, Ampho-rogynaceae,* and *Loranthaceae* go to back 80, 72, 53, 46, and 28 million years ago, respectively. In the 19th century, *Viscum album* and *Loranthaceae* were classified as part of the same family. However, Vidal-Rusell and Nick-rent [36] molecular analysis found that they come from different ancestors. There are three main varieties of *Viscum album*: (a) *Viscum album* subps. *alba*, which grows in hardwoods; (b) *Viscum album* subps. *abietis*, which grows in firs, both share obovate to oblong leaf shape; and (c) *Viscum album* subps. *austriacum*, which grows on pines and larches, and its leaf shape is lanceolate. *Viscum album* and *Loranthaceae* are the biggest families with 550 and 990 species, respectively.

*Viscum album* from Europe is the most studied plant. Extracts of *Viscum album* have been used as an alternative treatment and adjuvant cancer therapy in Germany, Austria, and Switzerland [27]. *Loranthaceae* family is mainly distributed in the tropical and subtropical regions of America [35]. The family is composed of three tribes: *Elytrantheae* (subtribes *Elytranthinae* and *Gaiadendrinae*), *Nuytsiae*, and *Lorantheae* (subtribes *Loranthinae* and *Psittacanthinae*) [35]. Karyological information, biogeography, and inflorescence/floral morphology were considered for these classifications.

*Viscaceae* family is found in America. Kenaley and Mathiasen [37] carried out MANOVA analysis to compare *Arceuthobium gillii* and *A. nigrum*, plants found in northern Mexico, and southern Arizona, which are known as dwarf mistletoe. This analysis demonstrates that both species are different among each other, the dimensions of the third internode and basal diameter of female and male plants were the characteristics in which both species are mainly different among each other. Significant differences in height, as well as flower and fruit size, were evident as well [37]. Dwarf mistletoe is the name used for a great variety of *Viscaceae* family. *Phoradendron brachystachyum* is another example of dwarf mistletoe. This plant is a hemiparasitic plant mostly distributed in Mexico, and it occupies many hectares of ponderosa pine not only in Mexico but also in the southern United States. *P. brachystachyum* grows in some zones of Mexico City and causes a negative impact on forest health, and its loss and waste. However, people of this region have used this plant as a diabetes treatment, this effect is attributed to morolic acid, and volatile compounds could be responsible for this effect [38, 39].

### 7.3.2   MISTLETOE LECTINS (MLS)

MLs cover a great variety of these glycoproteins. Table 7.2 shows a brief example of them. In the beginnings of MLs research, it was reported a toxic

**TABLE 7.2** Summary of Mistletoe Lectins and its General Characteristics

| Lectin Code | Source | Tree | Location | Sugar Affinity | Optimum pH | Temperature | References |
|---|---|---|---|---|---|---|---|
| HmRIP-RIP type II | V. album L. | Pyrus pashia-Wild apple | North-Western Himalaya (India) | L-Rhamnose, Meso-inositol, and L-Arabinose | 2.5–12.5 | 4–65°C | [11] |
| ML I and ML III | V. album var. Coloraturam | Quercus-Oak tree | Seoul, South Korea | D-galactose, Nacetyl-D-galactosamine, and lactose | 4.0–8.5 | 0–42°C | [41] |
| PpyLL | Not mentioned | Bauhinia monandra-Orchid tree | Brazil | Fructose, fructose-1-6-biphosphate | Up to 7.5 | Up to 75°C | [42] |
| VCA | Viscum album L var. Coloraturam | Quercus-Oak tree | Kangwon Province, Korea | D-galactose and N-acetylgalactosamine | Not mentioned | Not mentioned | [43] |
| ML II | Viscum album var. Coloratum | Quercus mongolica-Mongolian oak | Duk Yoo Mountain, located in Muju, Chollabukdo | D-Galactose and N-acetylgalactosamine | Under 9.0 | Not mentioned | [44] |
| Viscumin | V. album L. | Acer platanoides-Norway maple | Norway | Not mentioned | Under 8.0 | Not mentioned | [40] |

protein as poisonous properties of mistletoe. Olsness et al. [40] reported a protein called viscumin; this protein consisted of two chains around 29 and 34 kDa. They attributed the cytotoxicity effect to viscumin by means of inhibiting of cell-free protein synthesis. Nowadays, it is well known that mistletoe contains at least three types of lectins: ML I, II, and III, known as *Viscum album* agglutinins (VAAs) VAAs-I, II, and III as well. ML-I bind mainly to β-galactosides, ML-II to β-galactosides, and N-acetylgalactosides, whereas ML-III recognizes N-acetyl-galactosides. MLs induce intrinsic and extrinsic apoptosis in tumor cells by targeting distinct stages of the apoptotic pathways [28].

ML-I is the most studied lectin, and it consists of two subunits and belongs to the family of ribosomes, RIP type II. The subunit A inactivates the 60S ribosomal subunit of eukaryotic cells leading to the inhibition of protein synthesis, while the subunit B binds to galactose and saccharides. Subunit is responsible for binding to cell-surface glycoconjugates, which reveals the antitumoral effect of lectins [43, 44]. ML-L is coded in the 1923-nucleotide, sequenced in 1999 [46, 47]. Eck et al. cloned the mistletoe lectin gene and characterized it [46]. The authors report that subunit B, either recombinant or natural, has not a cytotoxicity effect, they concluded cytotoxicity only occurs when the protein with its subunit A is presented. Besides, their results suggest that the internalization of lectins is not dependent on glycosylation, although most of the studies attribute the internalization of lectin to glyco-sylation rate. The recombinant lectin obtained by Eck et al. does not present glycosylation due to it is not glycosylated in the process, concluding that the lack of glycosylation does not affect cytotoxicity activity [47].

It is suggested ML I is part of the defense system against insects, parasites, fungi, and bacteria. Meyer et al. [48] described a specific function of plant lectins. They described that the complex formed by ML-I and the growth hormone (phloretamids) inhibit the growth hormone of parasites, which lead to the death of parasites. ML-II is less studied and reported than ML-I, Yoon et al. [49] isolated a mistletoe lectin from Korean mistletoe (KM), a subspecies of *Viscum album*, grown on *Quercus mongolica*. The analysis proved that the lectin obtained was a lectin type II, with a molecular mass of 62.7 kDa in two subunits of 30.6 kDa and 32.5 kDa. KM (*Viscum album coloratum)* extract has a variety of biological activities, and it is specific to galactose and N-acetyl-galactosamine [45]. Korean mistletoe lectin (KML) is limitedly used in cancer therapy or as adjuvant due to its toxicity to normal cells [50]. KM plant is traditionally used as a sedative, antispasmodic, analgesic, and cardiotonic. These effects have been attributed to active compounds, such as lectins, steroids, triterpenes, sesquiterpene lactones, flavonoids, and alkaloids, which are part of the phytochemical composition of KM [13]. While European mistletoe (EM) has been extensively studied, KM is less explored as a therapeutic plant [51]. KML and EM lectin differ among each other in the molecular masses of subunits, isoelectric points, DNA sequences, and isotype forms [49]. EM lectins (VAAs) differs from KML in the molecular weights (Mw), VAAs have a molecular mass around 55 and 63 kDa, while KMLs have a molecular mass of 60 kDa. Authors discuss that the lower Mw may play an important role in the cytotoxicity activity; however, they observed a synergic effect of extract components and lectins, which lead to a higher cytotoxicity activity [52]. Yoon et al. [49] carried out a comparison among EM lectins and KMLs. They observed a lectin called KML-C consisting of four different chains in which some of the molecules are linked by disulfide bond; this lectin has affinity to GalNAc and D-Gal which suggest the possibility that KML-C is similar to ML type II [28] explain that the difference among MLs is owing to the host tree and the frost impact, which can be observed in the isoforms of lectins obtained. These differences are influenced by the interaction between mistletoe plants and their hosts, where the organic carbon transfer has a crucial influence [53].

As it has been mentioned, mistletoe is native from Europe, Asia, and North America, nevertheless, to the best of our knowledge, lectins from American mistletoe have been not reported before.

## 7.4  PHYSICOCHEMICAL PROPERTIES OF LECTINS

### 7.4.1  SUGAR BINDING SPECIFICITY

Taking advantage of the sugar-binding specificity, it is possible the use of lectins as a tool in a wide variety of processes. However, the main challenge is the necessity to reveal what type of carbohydrates specific to each lectin.

Although the exact mechanism is not known, it is assumed that interactions are due to electrostatic charge and non-polar forces. D-Mannose (D-Glucose), 2-Acetamido-2-deoxy-D-glucose, 2-Acetamido-2-deoxy-D-galactose, D-Galactose, and L-Fucose are some examples of sugars specifically bind with lectins [21].

Lectins bind a specific sequence of sugar moieties located in proteins or cell membranes containing these specific groups. Concanavalin A (ConA) is a lectin extracted from jack bean [22], which binds with membrane proteins glycosylated with a-D-glucose and a-D-mannose monomers [54].

Mannose and glucose are the most frequently reported sugars with which the mistletoe lectin can bind. Structures with mannose have been used to purify or immobilize different compounds trough this sugar-binding specificity [55]. Goncalves et al. [56] evaluated N-acetyl-D-glucosamine (D-GlcNAc) to purify lectin in affinity chromatography. Once they achieved the design of the purification system, they proceed to evaluate the specificity of lectins to various carbohydrates as ligands. N-acetyl-D-glucosamine, N-acetyl-D-mannosamine, and N-acetyl-D-galactosamine were evaluated with a higher factor of purification [57].

There are researches to determine the sequence of lectins. The results have demonstrated the presence of two QXDXNXVXY motifs in animal lectins that are involved in the binding of D-mannose to activate the immune response. Lectin of *C. striatus* shows a similarity of 50% in the genomic sequence when compared with the lectins of *Allium ampeloprasum* that have an affinity with the mannose [7].

Sugar binding specificity is related to the glycosylation process. The lectins can be classified according to two types of interactions involved in the specificity for sugar [58]: (a) N-glycosidic bond that occurs between the asparagine of the peptide chain and the OH group of carbohydrates. The process is carried out by the interaction of oligosaccharides such as 2 N-acetylglucosamine with a lectin residue asparagine. (b) Glucose, galactose, sialic acid, and the O-glycosidic bond in which the amino acids involved are serine and threonine. These oligosaccharides are known as glycans, which are oligomers of monosaccharide species that are linked together by several glycosidic linkages. This interaction must be studied for each lectin. In the case of MLs, the interaction with several sugars with great potential is known in a wide variety of applications.

### 7.4.2 METAL-BINDING

Lectins are considered as metalloid proteins. Their active structures are dependent on $Ca^{2+}$ and $Mn^{2+}$. Gondim et al. [59] evaluated ConBr, ConM, DSclerL, and DLasiL lectins. They confirmed the presence of $Mn^{2+}$, $Mg^{2+}$, and $Ca^{2+}$ significant amounts in four studied lectins. Metal ions such as $Ca^{2+}$ and $Mn^{2+}$ are important in plant lectins since divalent metals are required to create sugar-binding sites. Kaushik et al. [60] performed a simulation with a computational approach as molecular dynamics. This analysis allows the exploration of biological systems at the atomic level. They found that metal ions participate in the stabilization of the conformation of the key loop regions of ConA and allow the recognition of sugars. Although there are reports about the influence of metals on the functions of MLs, the content of the metal depends on the source and type of the mistletoe plant. Thus, the mineral analysis could be a reference to determine the content and type of metals in MLs before their applications.

### 7.4.3   OPTIMAL PH RANGE AND THERMAL STABILITY

Optimal pH influences the stability and functionality of proteins. Different lectins are characterized by a particular optimum pH in a wide range of values, which is probably related to their original role in the plant [61]. The lectin of *D. volubilis* has a wide range of optimum pH of 4.0 to 7.0 and good stability at 45°C [62]. This temperature can be favorable in the extraction process to achieve a higher extraction yield. There are other reports of lectins with an optimum pH range of 4.0 to 7.0, which are stable in a temperature range of 30 to 40C [41].

### 7.5   PURIFICATION OF LECTINS

### 7.5.1   CHROMATOGRAPHY

Affinity chromatography is a technique that has been used for lectin purification for a long time. The selective interaction with certain types of receptors is the most important characteristic of lectins. This interaction has been used to design the supports for affinity chromatography [63], which is a method that is currently applied for their purification. Affinity chromatography is a very efficient purification procedure, but it requires the pre-purification of samples to eliminate the column obstruction and increase the backpressure [64].

Some of the first reports of this technique described the application of support of Sepharosa® 6 B functionalized with N-acetylgalactosamine and N-acetylglucosamine for the purification of soybean and wheat lectins, with a maximum recovery capacity of 145 and 165 mg per each 100 g of grain, respectively [65].

Pohleven et al. [64] tested different conditions for lectin affinity chromatography. It was emphasized that due to the specificity of lectins for each type of glycan in different cells and microorganisms, it is not possible to standardize a carrier with broad coverage for many lectins. Proper selection of support, ligand, and elution techniques are crucial points to design a process that uses this technique. However, the appropriate conditions, carefully selected, could allow a successful selective separation [64].

Currently, there are reports of selective lectin purification by affinity chromatography with N-glycans [57] and monosaccharides such as mannose [66]. The type of lectin to be purified is an important factor for the development of the affinity chromatography method, whose selectivity depends on the selection of the ligand to ensure a better purification.

In addition to affinity chromatography, other types of chromatography have also been used for the purification of lectins. The lectin of *Laetiporus sulfureus* was purified by gel filtration chromatography [67]. Some other methods for purifying lectins are presented in Table 7.3.

**TABLE 7.3**  Methods Applied for Different Lectin Extraction

| Acronym (Organism) | Purification Method | References |
|---|---|---|
| LSL (*Laetiporus sulfureus*) | Affinity chromatography | [68] |
| DBL (*Dioscorea bulbifera*) | Affinity chromatography | [69] |
| AAL (*Aleuria aurantia*) | Affinity chromatography | [70] |
| LSL (*Laetiporus sulfureus*) | Salting precipitation | [67] |
| GaBL (*G. americana*) | Salting precipitation | [71] |
| AAL (*Aleuria Aurantia Lectin*) | Ion exchange chromatography | [72] |

### 7.5.2  MAGNETIC SEPARATION AS AN ALTERNATIVE OPTION TO PURIFY LECTINS

One strategy that has proved to be an excellent alternative consists of the application of magnetic separation for lectin purification. In this case, the selective ligands are applied to the magnetic nanoparticles. Nanoparticles have been used for the purification of proteins like enzymes [73–76]. For the specific case of the lectins, Heebøll-Nielsen, Dalkiær, and Thomas [77] reported procedure for the partial purification of the lectin from Jack Bean extract. The lectin has an affinity for mannose, so the nanoparticle has been functionalized with this ligand. Upon contact with the tuber extract, the lectin was adsorbed on the surface of the nanoparticles. The specific activity of the separated lectin increased 13 times after the purification process, and the recovery was 200 mg g$^{-1}$.

Recent studies have reported the use of mannose as a ligand for the selective purification of lectins, thus developing a synthesis method for these nanoparticles [55].

Moreover, gold nanoparticles have been used for the recovery of lectins with good results [78]. Considering the information previously mentioned, it is possible to suppose that the purification of lectins related to N-glycans can be carried out by magnetic separation in nanostructured support functionalized with the affinity ligand. The challenge will be to find the specific ligand for the lectin of interest.

Gregorio et al. [79] obtained a system of magnetic nanoparticles functionalized with the SA-α-2,6-Gal receptors of the avian influenza virus and with the N-SA-α-2,3-Gal receptor of the human influenza virus, both were extracted from the swine trachea tissue. These magnetic nanosystems interacted with the lectin of *Sambucus niger* and lectin *Maackia amurensis*, respectively. In order to extract the receptors from the swine trachea, both lectins were immobilized covalently in the chitosan-coated magnetite nanoparticles.

Magnetite nanoparticles have been widely used in recent years. One of the main problems of this type of nanoparticles is the formation of agglomerates in the presence and even in the absence of a magnetic field. This agglomeration increases with the coercivity and the remanence of the nanoparticles, thus losing their magnetic properties [80]. Moreover, magnetite nanoparticles are not resistant to strong acids or oxidizing agents that lead to the loss of magnetic activity [81].

## 7.5.3   MAGNETIC SEPARATION WITH CORE-SHELL NANOPARTICLES AS AN ALTERNATIVE OPTION TO PURIFY LECTINS

The core-shell technique avoids the formation of nanoparticle aggregates, maintaining their magnetic properties. Moreover, the use of core-shell allows to diminish or eliminate the possible toxicity of the nanoparticles. The choice of an inert material for their coating will be the key to guarantee safety. One of the materials used to carry out the core-shell are silicates, specifically tetraethyl orthosilicate (TEOS). The use of this coating allows the nanoparticles to stabilize to maintain their magnetic activity, in addition to the fact that it has very low toxicity [35, 36]. Nanoparticles with core-shells have been used in different biological applications: for the determination of endocrine disruptors such as 17β-estradiol and estriol, immobilization of myoglobin [82], among other applications.

## 7.6    MISTLETOE LECTIN APPLICATIONS

### 7.6.1    INFLAMMATORY MODULATION

MLs are immunomodulatory agents [13]. In general, plants have a wide variety of compounds with the activity of immune modulators. The lectins affect agonists of toll-like receptors (TLRs or TLRs), which participate in the detection of microbes and trigger inflammatory responses [83]. The KML is one of the main *Viscum album* compounds. KML increases the proliferation of T lymphocytes and B-lymphocytes, splenic NK cells, and macrophage activities *in vitro*. This response increases the production of cytokines such as interleukin-1 and interleukin-6 by macrophages, which helps to improve the activity of the immune system [13]. Schötterl et al. [84] examined the enhancing effects of three drugs containing ML-1 over the immune system. They observed an improvement of the immune system concluding that drugs are an excellent adjuvant in radiochemotherapy.

### 7.6.2    ANTICOAGULANT ACTIVITY

Homeostasis is the state of equilibrium in which the body is in physical and chemical conditions. Blood coagulation is an essential mechanism against bleeding. The leaf lectin of *Phthirusa pyrifolia* (PpyLL) is produced in a variety of mistletoe. Costa et al. [14] performed an analysis of the effect of radiation on the anticoagulant activity of PpyLL. It was noted that PpyLL could retain its anticoagulant activity at least up to 0.5 kGy. It was assumed that the PpyLL anticoagulant activity could be improved with radiation and used as a treatment in diseases in which homeostasis is altered. However, the anticoagulant activity must be studied to understand clearly how this is possible.

### 7.6.3    ANTITUMORAL ACTIVITY

Apoptosis is one of the mechanisms proposed to explain the effect of lectins on cancer cells. Apoptosis is programmed cell death, which is an active process related to a gradual modification in the morphology of cells. In general, fragments of DNA and chromatin are condensed, and then the nucleus is fragmented, which leads to cell fragmentation [24]. Thus, Con A (*Canavalia ensiformis* lectin) induces apoptosis in macrophage cells PU%-1.8 and A375 melanoma cells [85]. In cancer cells, the cell membrane is altered, showing residues of N-glycans, these compounds are known as Tn antigen. The Tn antigen refers to the structure of the monosaccharides N-acetylgalactosamine (GalNAc) linked to serine or threonine by a glycoside bond [86]. These oligosaccharides can be recognized by the lectins, which act by means of them [87].

The interest in lectins that interact with GalNAc is due to its effectiveness in decreasing the viability of cancer cells [9, 86]. Moreover, GalNAc forms part of the membranes of tumoral cells, making it a transcendental research object. It is known that membrane-bound glycoproteins (T and Tn antigens, for example, mucin) and glycolipids (Lewis a, Lewis x and Forssman antigens) are found on the surface of cancer cells. Galactose monosaccharide addition forms a disaccharide antigen known as the Thomsen-Friedenreich antigen (Gal (b1-3) GalNAc). The sialyl antigen Tn (STn antigen) is formed by elongation with sialic acid (Neu5Ac (a2-6) GalNAc) instead of galactose [87]. All these O-glycans have been identified as glyco-markers of interest for the diagnosis and prognosis of cancer diseases.

These epitopes are well detected using the specific lectins isolated from different plants, such as the mistletoe from the *Viscum album* [7, 87], yakalin of *Artocarpus integrifola* [88] and from fungi, such as the lectin of *Laetiporus sulfureus* [21, 89]. The T / Tn specific lectins are widely used for the histochemical detection of cancer cells in biopsies and to monitor the development of cancer. Specific T/Tn lectins also induce apoptosis in some cancer cell lines. Moreover, they can be used in the photodynamic treatment of tumors [90].

It has been shown that T/Tn antigens related to mucin are overexpressed on the surface of cancer cells, providing a specific target for cancer immunotherapy [86]. Parry et al. [91] designed the multivalent glycosylated peptide-free system as possible synthetic vaccines against cancer. The application of this system *in vivo* assays led to a significant increase of antibodies titers selective to mucin glycans. In the design of the system, gold nanoparticles and a glycopolymer obtained by polymerization of Tn antigen were used.

## 7.7    CONCLUSION

Lectin has demonstrated great potential due to the ability to recognize the glycosyl receptors of target cells. This selective interaction makes lectin from *Viscum album* (mistletoe) lectins an essential tool for the diagnosis of diseases, such as cancer, as well as their treatment. However, this sugar-binding specificity represents a significant challenge due to each lectin has a specific way to join with sugars. Besides to be a challenge, lectins have a wide array of biological applications. However, most of these applications can be studied further to the complete understanding of lectins mechanism. Therefore, lectins can be considered as a potential source of highly biotechnological products. Mistletoe plant represents an excellent opportunity to research, showing them as a promising plant due to its phytochemical compound discovered and undiscovered.

## KEYWORDS

- **affinity chromatography**
- **anticancer effect**
- **European mistletoe**
- **Korean mistletoe lectin**
- **lectins**
- **mistletoe**
- **nanoparticles**

## REFERENCES

1. Hajto, T., Krisztina, F., Ildiko, A., Zsolt, P., Peter, B., Peter, N., & Pal, P., (2007). Unexpected different binding of mistletoe lectins from plant extracts to immobilized lactose and N-acetylgalactosamine. *Analytical Chemistry Insights, 2*, 43–50. Retrieved from: http://www.ncbi.nlm.nih.gov/entrez/query.fcgi?cmd=Retrieve&db=PubMed&dopt=Citation&list_uids=19662176 (accessed on 22 December 2020).
2. Anderson, R. K. G. K., & Evers, D., (2008). Structure and function of mammalian carbohydrate-lectin interactions. *Glycoscience*, 2445–2482. https://doi.org/10.1007/978-3-540-30429-6_63.
3. Tasumi, S., Yamaguchi, A., Matsunaga, R., Fukushi, K., & Suzuki, Y., Nakamura, O., & Kikuchi, K., Tsutsui, S., (2016). Identification and characterization of pufflectin from the grass pufferfish *Takifugu niphobles* and comparison of its expression with that of

*Takifugu rubripes. Developmental and Comparative Immunology, 59*, 48–56. https://doi.org/10.1016/j.dci.2016.01.007.

4. Loh, S. H., Park, J. Y., Cho, E. H., Nah, S. Y., & Kang, Y. S., (2017). Animal lectins: Potential receptors for ginseng polysaccharides. *Journal of Ginseng Research, 41*(1), 1–9. http://doi.org/10.1016/j.jgr.2015.12.006.

5. Kilpatrick, D. C., (2002). Animal lectins: A historical introduction and overview. *Biochimica et Biophysica Acta-General Subjects, 1572*(2/3), 187–197.

6. Tsutsui, S., Tasumi, S., Suetake, H., & Suzuki, Y., (2003). Lectins homologous to those of monocotyledonous plants in the skin mucus and intestine of pufferfish (*Takifugu rubripes*). *Journal of Biological Chemistry, 278*(23) 20882–20889. http://doi.org/10.1074/jbc.M301038200.

7. Arasu, A., Arasua, A., & Kumaresana, V., Sathyamoorthia, A., Palanisamya, R., Prabhaa, N., Bhatta, P., et al., (2013). Fish lily type lectin-1 contains β-prism architecture: Immunological characterization. *Molecular Immunology, 56*(4), 497–506. http://doi.org/10.1016/j.molimm.2013.06.020.

8. Yin, X., Mu, L., Li, Y., Wu, L., Yang, Y., Bian, X., Li, B., et al., (2019). Identification and characterization of a B-type mannose-binding lectin from Nile tilapia (*Oreochromis niloticus*) in response to bacterial infection. *Fish Shellfish Immunology, 84*, 91–99. https://doi.org/10.1016/j.fsi.2018.09.072.

9. Golotin, V. A., Filshtein, A. P., Chikalovets, I. V., Yu, K. N., Molchanova, V. I., & Chernikov, O. V., (2019). Expression and purification of a new lectin from mussel *Mytilus trossulus. Protein Expression and Purification, 154,* 62–65. https://doi.org/10.1016/j.pep.2018.10.003.

10. Sánchez-Salgado, J. L., Pereyraa, M. A., Agundisa, C., Vivanco-Rojasa, O., Rosalesc, C., Pascual, C., Alpuche-Osornoe, J. J., & Zentenoa, E., (2018). The effect of the lectin from *Cherax quadricarinatus* on its granular hemocytes. *Fish Shellfish Immunology, 77*, 131–138. https://doi.org/10.1016/j.fsi.2018.03.050.

11. Aranda-Souza, M. Â., de Lorena, V. M. B., Dos, S. C. M. T., & De Figueiredo, R. C. B. Q., (2018). *In vitro* effect of *Bothrops leucurus* lectin (BLL) against *Leishmania amazonensis* and *Leishmania braziliensis* infection. *International Journal of Biological Macromolecules, 120*, 431–439. https://doi.org/10.1016/j.ijbiomac.2018.08.064.

12. Huang, Z., Ma, A., & Xia, D., Wang, X., Sun, Z., Shang, X., Yang, Z., & Qu, J., (2016). Immunological characterization and expression of lily-type lectin in response to environmental stress in turbot (*Scophthalmus Maximus*). *Fish Shellfish Immunology, 58*, 323–331. http://doi.org/10.1016/j.fsi.2016.08.025.

13. Lee, C. H., Kim, J. K., Kim, H. Y., Park, S. M., & Lee, S. M., (2009). Immunomodulating effects of Korean mistletoe lectin *in vitro* and *in vivo. International Immunopharmacology, 9*(13/14), 1555–1561. http://doi.org/10.1016/j.intimp.2009.09.011.

14. Costa, R., Albuquerque, W. W., Campos-Silva, M. C. C., De Melo, P. R. A., Mychely, S. A., Porto, M. L. V., & Figueiredo, A. L., (2017). Can γ-radiation modulate hemagglutinating and anticoagulant activities of PpyLL, a lectin from *Phthirusa pyrifolia? International Journal of Biological Macromolecules, 104*, 125–136. http://doi.org/10.1016/j.ijbiomac.2017.06.007.

15. Melo, M. N. O., Oliveira, A. P., Wiecikowski, A. F., Carvalho, R. S., Castro, J. L., De Oliveira, F. A. G., Pereira, H. M. G., et al., (2018). Phenolic compounds from *Viscum album* tinctures enhanced antitumor activity in melanoma murine cancer cells. *Saudi Pharmaceutical Journal, 26*(3), 311–322. https://doi.org/10.1016/j.jsps.2018.01.011.

16. Kirsch, A., & Hajto, T., (2011). Case reports of sarcoma patients with optimized lectin-oriented mistletoe extract therapy. *The Journal of Alternative and Complementary Medicine, 17*(10), 973–979. https://doi.org/10.1089/acm.2010.0596.

17. Pervin, M., Koyama, Y., & Isemura, M., (2015). Plant lectins in therapeutic and diagnostic cancer research. *International Journal of Plant Biology and Research, 3*(2), 1030–1035.

18. Liener, I. E., Sharon, N., & Goldstein, I. J., (1986). *The Lectins: Properties, Functions, and Applications in Biology and Medicine.* Academic Press. https://doi.org/10.1016/B978-0-12-449945-4.X5001-5.

19. Goldstein, I. J., & Hayes, C. E., (1978). The lectins: Carbohydrate-binding proteins of plants and animals. *Advances in Carbohydrate Chemistry and Biochemistry, 35,* 127–340. https://doi.org/10.1016/S0065-2318(08)60220-6.

20. Da Silva, T. A., Mariano, V. S., Sardinha-Silva, A., De Souza, M. A., Mineo, T. W. P., & Roque-Barreira, C. M., (2016). IL-17 induction by ArtinM is due to stimulation of IL-23 and IL-1 release and/or interaction with CD3 in CD4+ T-cells. *PLoS One, 11*(2), 1–16. https://doi.org/10.1371/journal.pone.0149721.

21. Cagliari, R., Kremer, F. S., & Pinto, L. S., (2018). Bauhinia lectins: Biochemical properties and biotechnological applications. *International Journal of Biological Macromolecules, 119,* 811–820. https://doi.org/10.1016/j.ijbiomac.2018.07.156.

22. Ríos-De, Á. L., Jackson, F., Greer, A., Bartley, Y., Bartley, D. J., Grantd, G., & Huntley, J. F., (2012). *In vitro* screening of plant lectins and tropical plant extracts for anthelmintic properties. *Veterinary Parasitology, 186*(3/4), 390–398. http://doi.org/10.1016/j.vetpar.2011.11.004.

23. De Freitas-Pires, A., Marques-Bezerra, M., Ferreira-Amorim, R. M., Do Nascimento, F. L. M., Marinho, M. M., Mata-Moura, R., Lima-Silva, M. T., et al., (2018). PII, LECTIN purified from *Lonchocarpus campestris* seeds inhibits inflammatory nociception. *International Journal of Biological Macromolecules, 125,* 53–60. http://doi.org/10.1016/j.ijbiomac.2018.11.233.

24. Liu, B., Bian, J. H., & Bao, J., (2010). Plant lectins: Potential antineoplastic drugs from bench to clinic. *Cancer Letters, 287*(1), 1–12. http://doi.org/10.1016/j.canlet.2009.05.013.

25. Nascimento, A. P. M., Knaut, J. L., Rieger, D. K., Wolin, I. A. V., Heinrich, I. A., Mann, J., Juarez, A. V., et al., (2018). Anti-glioma properties of DVL, a lectin purified from dioclea violacea. *International Journal of Biological Macromolecules, 120,* 566–577. https://doi.org/10.1016/j.ijbiomac.2018.08.106.

26. Zárate, G., Sáez, G. D., & Pérez, A., (2017). Dairy propionibacteria prevent the proliferative effect of plant lectins on SW480 cells and protect the metabolic activity of the intestinal microbiota *in vitro. Anaerobe, 44,* 58–65. https://doi.org/10.1016/j.anaerobe.2017.01.012.

27. Zänker, K. S., & Kaveri, S. V., (2015). *Mistletoe: From Mythology to Evidence-Based Medicine* (1ˢᵗ edn.). Karger, Germany.

28. Arndt, B., (2004). *Mistletoe: The Genus Viscum* (1ˢᵗ edn.). Taylor & Francis e-Library, Herdecke, Germany.

29. Ochocka, J. R., & Piotrowski, A., (2002). Biologically active compounds from European mistletoe (*Viscum album* L.). *Canadian Journal of Plant Pathology, 24*(1), 21–28. https://doi.org/10.1080/07060660109506966.

30. Vidal-Russell, R., & Nickrent, D. L., (2008a). Evolutionary relationships in the showy mistletoe family (Loranthaceae). *American Journal of Botany, 95*(8), 1015–1029. https://doi.org/10.3732/ajb.0800085.

31. Westwood, J. H., Yoder, J. I., Timko, M. P., & Claude, W., (2010). The evolution of parasitism in plants. *Trends in Plant Science, 15*(4), 227–235. http://doi.org/10.1016/j.tplants.2010.01.004.

32. Bar-Sela, G., (2011). White-berry mistletoe (*Viscum album* L.) as complementary treatment in cancer: Does it help? *European Journal of Integrative Medicine, 3*(2), e55–e62. https://doi.org/ 10.1016/j.eujim.2011.03.002.

33. Liu, B., Le, C. T., Barrett, R. L., Nickrent, D. L., Chen, Z., Lu, L., & Vidal-Russell, R., (2018). Historical biogeography of *Loranthaceae* (Santalales): Diversification agrees with the emergence of tropical forests and radiation of songbirds. *Molecular Phylogenetics and Evolution, 124*, 199–212. https://doi.org/10.1016/j.ympev.2018.03.010.

34. Wu, D. Y., Raven, Z. Y., & Hong, P. H., (2003). *Viscaceae P. in Flora of China (Ulmaceae through Basellaceae).* Beijing and Missouri Botanical Garden Press, Science Press.

35. Vidal-Russell, R., & Nickrent, D. L., (2007). The biogeographic history of Loranthaceae. *Darwiniana, 45*, 52–54.

36. Vidal-Russell, R., & Nickrent, D. L., (2008b). The first mistletoes: origins of aerial parasitism in santalales. *Molecular Phylogenetics and Evolution, 47*(2), 523–537. https://doi.org/ 10.1016/j.ympev.2008.01.016.

37. Kenaley, S. C., & Mathiasen, R. L., (2014). *Arceuthobium gillii* and A. nigrum (viscaceae) revisited: Distribution, morphology, and RDNA-its analysis. *Journal of the Botanical Research Institute of Texas, 7*(1), 311–322. https://www.jstor.org/stable/24621076.

38. López-Martínez, S., Navarrete-Vázquez, G., León-Rivera, I., & Rios, M. Y., (2013). Chemical constituents of the hemiparasitic plant *Phoradendron* brachystachyum DC Nutt (Viscaceae). *Natural Product Research, 27*(2), 130–136. https://doi.org/10.1080/ 14786419.2012.662646.

39. Hoyt, H. M., Hornsby, W., Huang, C. H., Jacobs, J. J., & Mathiasen, R. L., (2017). Dwarf mistletoe control on the mescalero apache Indian reservation, New Mexico. *Journal of Forestry, 115*(5), 379–384. https://doi.org/10.5849/jof.16–049.

40. Olsness, A., Stirpe, S., Sandvig, F., & Phil, K., (1982). Isolation and characterization of viscumin, a toxic lectin from *Viscum album* L. (Mistletoe). *Journal of Biological Chemistry, 257*, 13263–13270.

41. Park, W. B., Han, S. K., Lee, M. H., & Han, K. H., (1997). Isolation and characterization of lectins from stem and leaves of Korean mistletoe (Viscum album var. coloratum) by affinity chromatography. *Archives of Pharmacal. Research, 20*(4), 306–312. https://doi.org/10.1007/BF02976191.

42. Costa, R. M. P. B., Vaz, F. M., Oliva, M. L. V., Coelho, C. B. B. L., Correia, M. T. S., & Carneiro-Da-Cunha, M. G., (2010). A new mistletoe *Phthirusa* pyrifolia leaf lectin with antimicrobial properties. *Process Biochemistry Journal, 45*, 526–533. https://doi.org/ 10.1016/j.procbio.2009.11.013.

43. Lyu, S. Y., Park, S. M., Choung, B. Y., & Park, W. B., (2000). Comparative study of Korean (*Viscum album* var. coloratum) and European mistletoes (*Viscum album*). *Medicinal Chemistry and Natural Product, 23*(6), 592–598. https://doi.org/10.1007/BF02975247.

44. Lee, H. S., Kim, Y. S., Kim, S. B., Choi, B. E., Woo, B. H., & Lee, K. C., (1999). Isolation and characterization of biologically active lectin from Korean mistletoe,

viscum album var. coloratum. *Cellular and Molecular Life Science, 55*, 679–682. https://doi.org/10.1007/s000180050324.

45. Kim, J. J., Hwang, Y. H., Kang, K. Y., Kim, I., Kim, J. B., Park, J. H., Yoo, Y. C., & Yee, S. T., (2014). Enhanced dendritic cell maturation by the B-chain of Korean mistletoe lectin (KML-B), a novel TLR4 agonist. *International Immunopharmacology, 21*(2), 309–319. http://dx.doi.org/10.1016/j.intimp.2014.05.010.

46. Eck, J., Langer, M., Mo, B., Baur, A., Rothe, M., Zinke, H., & Lentzen, H., (1999a). Cloning of the mistletoe lectin gene and characterization of the recombinant A-chain. *European Journal of Biochemistry*, 775–784.

47. Eck, J., Langer, M., Mo, B., Witthohn, K., Zinke, H., & Lentzen, H., (1999b). Characterization of recombinant and plant-derived mistletoe lectin and their B-chains. *European Journal of Biochemistry, 797,* 788–797. https://doi.org/10.1046/j.1432-1327.1999.00784.x.

48. Meyer, A., Rypniewski, W., Celewicz, L., Erdmann, V. A., Voelter, W., Singh, T. P., Genov, N., et al., (2007). The mistletoe lectin I-phloretamide structure reveals a new function of plant lectins. *Biochemical and Biophysical Research Communications, 364*, 195–200. https://doi.org/10.1016/j.bbrc.2007.09.113.

49. Yoon, T. J., Yoo, Y. C., Kang, T. B., Shimazaki, K., Song, S. K., Leea, K. H., Kima, S. H., et al., (1999). Lectins isolated from Korean mistletoe (*Viscum album* coloratum) induce apoptosis in tumor cells. *Cancer Letters, 136*, 33–40. https://doi.org/10.1016/S0304-3835(98)00300-0.

50. Lyu, S. Y., & Park, W. B., (2010). Mistletoe lectin transport by M-cells in follicle-associated epithelium (FAE) and IL-12 secretion in dendritic cells situated below FAE *in vitro. Archives of Pharmacal Research, 33*(9), 1433–1441. https://doi.org/ 10.1007/s12272-010-0918-6.

51. Lyu, S. Y., Choi, J. H., Lee, H. J., Park, W. B., & Kim, G. J., (2013). Korean mistletoe lectin promotes proliferation and invasion of trophoblast cells through regulation of Akt signaling. *Reproductive Toxicology, 39*, 33–39. http://dx.doi.org/10.1016/j.reprotox.2013.03.011.

52. Lyu, S., Kwon, Y., Joo, H., & Park, W., (2004). Preparation of alginate/chitosan microcapsules and enteric-coated granules of mistletoe lectin. *Archives of Pharmacal. Research, 27*(1), 118–126. https://doi.org/10.1007/bf02980057.

53. Těšitel, J., Plavcová, L., & Cameron, D. D., (2010). Interactions between hemiparasitic plants and their hosts: The importance of organic carbon transfer. *Plant Signaling and Behavior, 5*(9), 1072–1076. https://doi.org/10.4161/psb.5.9.12563.

54. Bhattacharyya, L., Ceccarini, C., Lorenzoni, P., & Brewer, C. F., (1987). Concanavalin an interactions with asparagine-linked glycopeptides. *Journal of Biological Chemistry, 262*(3), 1288–1293.

55. Demir, F. E., Atay, O. N., Koruyucu, M., Kök, G., Salman, Y., & Akgöl, S., (2018). Mannose based polymeric nanoparticles for lectin separation. *Separation Science and Technology, 53*(5), 2365–2375. https://doi.org/10.1080/01496395.2018.1452943.

56. Gonçalves, G. R. F., Gandolfi, O. R. R., Santos, C. M. S., Bonomo, R. C. F., Veloso, C. M., & Fontan, R. C. I., (2016). Development of supermacroporous monolithic adsorbents for purifying lectins by affinity with sugars. *Journal of Chromatography B: Analytical Technologies in the Biomedical and Life Sciences, 1033–1034*, 406–412. https://doi.org/10.1016/j.jchromb.2016.09.016.

57. Gonçalves, G. R. F., Gandolfi, O. R. R., Santos, L. S., Bonomo, R. C. F., Veloso, C. M., Cristiane, M., Veríssimo, L. A. A., & Fontan, R. C. I., (2017). Immobilization of sugars in supermacroporous cryogels for the purification of lectins by affinity chromatography. *Journal of Chromatography B: Analytical Technologies in the Biomedical and Life Sciences, 1068–1069*, 71–77. https://doi.org/10.1016/j.jchromb.2017.10.019.

58. Zhang, P., Woen, S., Wang, T., Liau, B., Zhao, S., Chen, C., Yang, Y., et al., (2016). Challenges of glycosylation analysis and control: An integrated approach to producing optimal and consistent therapeutic drugs. *Drug Discovery Today, 21*(5), 740–765. https://doi.org/10.1016/j.drudis.2016.01.006.

59. Gondim, A. C. S., Romero-Canelón, I., Sousa, E. H. S., Blindauer, C. A., Butler, J. S., Romero, M. J., Sanchez-Cano, C., et al., (2017). The potent anti-cancer activity of dioclea lasiocarpa lectin. *Journal of Inorganic Biochemistry, 175,* 179–189. https://doi.org/10.1016/j.jinorgbio.2017.07.011.

60. Kaushik, S., Mohanty, D., & Surolia, A., (2009). The role of metal Ions in substrate recognition and stability of concanavalin A: A molecular dynamics study. *Biophysical Journal, 96*(1), 21–34. https://doi.org/10.1016/j.jinorgbio.2017.07.011.

61. Talley, K., & Alexov, E., (2013). On the pH-optimum of activity and stability of proteins Kemper. *Proteins, 78*(12), 2699–2706. https://doi.org/10.1038/mp.2011.182.doi.

62. Patil, M. B., & Deshpande, K. V., (2015). Isolation and characterization of lectin from leaves of *Dregea volubilis. Journal of Global Biosciences, 4*(6), 2496–2503.

63. Freeze, H., (1995). Lectin affinity chromatography. *Current Protocols in Protein Science, 00*(1), 9.1.1–9.1.9. https://doi.org/10.1002/0471140864.ps0901s00.

64. Pohleven, J., Trukelj, B., & Kos, J., (2012). In: Magdeldin, S., (ed.), *Affinity Chromatography of Lectins in Affinity Chromatography* (pp. 50–73). Slovenia. https://doi.org/10.5772/36578.

65. Vretblad, P., (1976). Purification of lectins by biospecific affinity chromatography. *Biochimica et Biophysica Acta, 434,* 169–176. https://doi.org/10.1016/0005-2795(76)90047-7.

66. Totten, S. M., Adusumilli, R., Kullolli, M., Tanimoto, C., Brooks, J. D., Mallick, P., & Pitteri, S. J., (2018). Multi-lectin affinity chromatography and quantitative proteomic analysis reveal differential glycoform levels between prostate cancer and benign prostatic hyperplasia sera. *Scientific Reports, 8*(1), 1–13. https://doi.org/ 10.1038/s41598-018-24270-w.

67. Wang, Y., Wu, B., Shao, J., Jia, J., Tian, Y., Shu, X., Ren, X., & Guan, Y., (2018). Extraction, purification, and physicochemical properties of a novel lectin from *Laetiporus sulphureus* mushroom. *LWT-Food Science and Technology, 91,* 151–159. https://doi.org/10.1016/j.lwt.2018.01.032.

68. Li, X. J., Liu, J. L., Gao, D. S., Wan, W., Yang, X., Li, Y. T., Chang, H. T., et al., (2016). Single-step affinity and cost-effective purification of recombinant proteins using the sepharose-binding lectin-tag from the mushroom *Laetiporus sulphureus* as fusion partner. *Protein Expression and Purification, 119,* 51–56. http://doi.org/10.1016/j.pep.2015.11.004.

69. Sharma, M., Hotpet, V., Sindhura, B. R., Kamalanathan, A. S., Swamy, B. M., & Inamdar, S. R., (2017). Purification, characterization, and biological significance of mannose binding lectin from *Dioscorea bulbifera* bulbils. *International Journal of Biological Macromolecules, 102,* 1146–1155.

70. Bergström, M., Åström, E., Påhlsson, P., & Ohlson, S., (2012). Elucidating the selectivity of recombinant forms of *Aleuria aurantia* lectin using weak affinity chromatography. *Journal of Chromatography B: Analytical Technologies in the Biomedical and Life Sciences, 885, 886*, 66–72. http://doi.org/10.1016/j.jchromb.2011.12.015.

71. Costa, R. B., Campana, P. T., Chambergo, F. S., Napoleão, T. H., Paiva, P. M., Guedes, P. P. M., Pereira, H. J. V., et al., (2018). Purification and characterization of a lectin with refolding ability from *Genipa americana* bark. *International Journal of Biological Macromolecules, 119*, 517–523. https://doi.org/10.1016/j.ijbiomac.2018.07.178.

72. Cai, D., Xun, C., Tang, F., Tian, X., Yang, L., Ding, K., Li, W., et al., (2017). Glycoconjugate probes containing a core-fucosylated N-glycan trisaccharide for fucose lectin identification and purification. *Carbohydrate Research, 449*, 143–152. http://doi.org/10.1016/j.carres.2017.07.011.

73. Mirzajani, F., Motevalli, S. M., Jabbari, S., Ranaei, S. S. O., & Sefidbakht, Y., (2017). Recombinant Acetylcholinesterase purification and its interaction with silver nanoparticle. *Protein Expression and Purification, 136*, 58–65. https://doi.org/10.1016/j.pep.2017.05.007.

74. Zengin, K. B., Uckaya, F., & Durmus, Z., (2017). Chitosan and carboxymethyl cellulose-based magnetic nanocomposites for application of peroxidase purification. *International Journal of Biological Macromolecules, 96*, 149–160. http://doi.org/10.1016/j.ijbiomac.2016.12.042.

75. Pérez-Guzmán, A. K., García-García, J. D., Ilyina, A., Zugasti-Cruz, A., Segura-Ceniceros, E. P., Martínez-Hernández, J. L., & Ramos-Gonzalez, R., (2019). Chitosan-heparin functionalized magnetic nanoparticles for the magnetic recovery of *Aspergillus niger* lipase enzyme. *Micro and Nano Letters, 14*(6), 623–628. https://doi.org/10.1049/mnl.2018.5490.

76. Carneiro, L. A. B. C., & Ward, R. J., (2018). Functionalization of paramagnetic nanoparticles for protein immobilization and purification. *Analytical Biochemistry, 540–541*, 45–51. https://doi.org/10.1016/j.ab.2017.11.005.

77. Heebøll-Nielsen, A., Dalkiær, M., & Thomas, O. R. T., (2004). Superparamagnetic adsorbents for high-gradient magnetic fishing of lectins out of legume extracts. *Biotechnology and Bioengineering, 87*, 311–323. https://doi.org/10.1002/bit.20116.

78. Selvaprakash, K., & Chen, Y. C., (2018). Functionalized gold nanoparticles as affinity nanoprobes for multiple lectins. *Colloids and Surfaces B: Biointerfaces, 162*, 60–68. http://doi.org/10.1016/j.colsurfb.2017.11.022.

79. Gregorio-Jauregui, K. M., Carrizalez-Alvarez, S. A., Rivera-Salinas, J. E., Saade, H., Martinez, J. L., López, R. G., Segura, E. P., & Ilyina, A., (2014). Extraction and Immobilization of SA-α-2, 6-Gal Receptors on magnetic nanoparticles to study receptor stability and interaction with *Sambucus nigra* lectin. *Applied Biochemistry and Biotechnology, 172*, 3271–3735. https://doi.org/10.1007/s12010-014-0801-x.

80. Barnakov, Y. A., Yu, M. H., & Rosenzweig, Z., (2005). Manipulation of the magnetic properties of magnetite-silica nanocomposite materials by controlled stober synthesis. *American Chemical Society, 11*, 7524–7527. https://doi.org/10.1021/la0508893.

81. Enache, D. F., Vasile, E., Simonescu, C. M., Răzvan, A., Nicolescu, A., Nechifor, A. C., Oprea, O., et al., (2017). Cysteine-functionalized silica-coated magnetite nanoparticles as potential nanoadsorbents. *Journal of Solid-State Chemistry, 253*, 318–328. http://doi.org/10.1016/j.jssc.2017.06.013.

82. Peng, H. P., Liang, R. P., Zhang, L., & Qiu, J. D., (2011). Sonochemical synthesis of magnetic core-shell Fe3O4@ZrO2nanoparticles and their application to the highly effective immobilization of myoglobin for direct electrochemistry. *Electrochimica Acta, 56*(11), 4231–4236. http://doi.org/10.1016/j.electacta.2011.01.090.

83. O'Neill, L. A. J., Hennessy, E. J., & Parker, A. E., (2010). Targeting Toll-like receptors: Emerging therapeutics? *Nature Reviews Drug Discovery, 9*(4), 293–307. https://doi.org/10.1038/nrd3203.

84. Schötterl, S., Huber, S. M., Lentzen, H., Mittelbronn, M., & Naumann, U., (2018). Adjuvant therapy using mistletoe-containing drugs boosts the T-cell-mediated killing of glioma cells and prolongs the survival of glioma bearing mice. *Evidence-Based Complementary and Alternative Medicine,* 1–12. https://doi.org/10.1155/2018/3928572.

85. Wu, L., Liu, T., Xiao, Y., Li, X., Zhu, Y., Zhao, Y., Bao, J., & Wu, C., (2015). Polygonatum odoratum lectin induces apoptosis and autophagy by regulation of microRNA-1290 and microRNA-15a-3p in human lung adenocarcinoma A549 cells. *International Journal of Biological Macromolecules, 85,* 217–226. http://dx.doi.org/10.1016/j.ijbiomac.2015.11.014.

86. Mortezai, N., Behnken, H. N., Kurze, A. K., Ludewig, P., Friedrich, B., Meyer, B., & Wagener, C., (2013). Tumor-associated Neu5Ac-Tn and Neu5Gc-Tn antigens bind to C-type lectin CLEC10A (CD301, MGL). *Glycobiology, 23*(7), 844–852. https://doi.org/10.1093/glycob/cwt021.

87. Fu, C., Zhao, H., Wang, Y., Cai, H., Xiao, Y., Zeng, Y., & Chen, H., (2016). Tumor-associated antigens: Tn antigen, sTn antigen, and T antigen. *HLA, 88*(6), 275–286. https://doi.org/10.1111/tan.12900.

88. Poiroux, G., Barre, A., Van, D. E. J. M., Benoist, H., & Rougé, P., (2017). Plant lectins targeting O-glycans at the cell surface as tools for cancer diagnosis, prognosis, and therapy. *International Journal of Molecular Science, 18,* 1232. https://doi.org/10.3390/ijms18061232.

89. Chang, Y. S., Chen, J. N., Chang, K. H., Chang, Y. M., Lai, Y. J., & Liu, W. J., (2019). Cloning and expression of the lectin gene from the mushroom *Agrocybe aegerita* and the activities of recombinant lectin in the resistance of shrimp white spot syndrome virus infection. *Developmental and Comparative Immunology, 90,* 1–9. https://doi.org/10.1016/j.dci.2018.07.020.

90. Magnadóttir, B., Bragason, B. T., Bricknell, I. R., Bowden, T., Nicholas, W. P., Hristova, M., Guðmundsdóttir, S., et al., (2019). Peptidyl arginine deiminase and deiminated proteins are detected throughout early halibut ontogeny-complement components C3 and C4 are post-translationally deiminated in halibut (*Hippoglossus hippoglossus* L.). *Developmental and Comparative Immunology, 92,* 1–19. https://doi.org/10.1016/j.dci.2018.10.016.

91. Parry, A. L., Clemson, N. A., Ellis, J., Bernhard, S. S. R., Davis, B. G., & Cameron, N. R., (2013). 'Multicopy multivalent' glycopolymer-stabilized gold nanoparticles as potential synthetic cancer vaccines. *Journal of the American Chemical Society, 135*(25), 9362–9365. https://doi.org/10.1021/ja4046857.

92. Mishra, V., Shyam, R., & Paramasivam, M., (2005). cDNA cloning and characterization of a ribosome-inactivating protein of a hemiparasitic plant (*Viscum album* L.) from North-Western Himalaya (India). *Plant Science, 168,* 615–625. https://doi.org/10.1016/j.plantsci.2004.09.024.

93. Nickrent, D. L., (2017). *Santalales (Including Mistletoes)* (pp. 1–6). Encyclopedia of Life Sciences, John Wiley & Sons. https://doi.org/10.1002/9780470015902.a0003714. pub2.

94. Peumans, W. J., & Vanm, D. E. J. M., (1995). The role of lectins in plant defense. *The Histochemical Journal, 27*, 253–271. https://doi.org/10.1007/BF00398968.

95. Vicas, S. I., & Socaciu, C., (2007). The biological activity of European mistletoe (*Viscum album*) extracts and their pharmaceutical impact. *Bulletin USAMV-CN, 23*(1120), 217–222. http://doi.org/10.15835/buasvmcn-agr:1344.

# CHAPTER 8

# Study of Refractance Window Dried Shredded Carrot and Analysis of Its Physical Properties

PREETISAGAR TALUKDAR,[1] PRANJAL PRATIM DAS,[1] and
MANUJ KUMAR HAZARIKA[2]

[1]Department of Chemical Engineering, IIT Guwahati, Assam, India,
E-mail: preetisagar1891@gmail.com (P. Talukdar)

[2]Department of Food Engineering and Technology, Tezpur University,
Assam, India

## ABSTRACT

The shredded carrot was dried in a laboratory set up of refractance window drying (RWD) at a temperature of 95°C for 75 minutes. The size of the shredded carrot was maintained in the range of 1.02–1.25 mm with a commercial vegetable cutter. Before drying, the carrot was blanched in hot water at 100°C for 6 minutes. In the experimental setup, radiating energy from a reservoir of hot water at an elevated temperature of 95°C was used for drying the vegetables. The moisture content of carrot, dried for the different period, was measured by the oven drying method, and drying analysis was done by the application of Fick's second law of diffusion and other thin layer models. The modified Henderson and Pabis model was found to be the best-fitted model with $R^2$ of the value 0.983 and RMSE at 0.03676. Quality parameters of dried carrots like moisture content, carotene content, color measurement, rehydration ratio, texture, and water activity were determined by standard experimental procedures. From the experimental measurements, it has been found that the drying time of the sample in the RWD system was less than that with other conventional drying methods. It was also seen that the method of RWD gave a better product, as was apparent from the experimentally measured values of carotene content (499.07 ± 3.78 µg).

## 8.1   INTRODUCTION

In the food industry, the drying of fresh foods was regarded as an important process. Compared to other preservation methods, drying offers a better quality product. Dried products were resistant to spoilage by microorganisms, have reduced water activity, reduced chemical reaction, can be stored for a long duration, and transported easily. In addition, as a result of changing lifestyles dried food demands were increasing [1, 2].

There were several drying techniques like drum drying, hot air oven drying, rotary drying, spray drying, tray drying, fluidized bed drying, freeze-drying, microwave drying, infrared drying [3]. Selecting a suitable method was difficult as not all methods give good results. Hot air oven drying and tray drying were cheap drying methods, but product quality was low while freeze-drying, microwave drying, and infrared drying gives better quality products but were expensive. Therefore, a drying technique was needed that offer better quality product and was cost-effective.

Refractance window (RW™) drying (RWD) developed by MCD Technologies, Inc. was a novel drying method that gives better-quality product by drying at high temperature in a short time [4]. In this system, thermal heat from hot water (95–97°C) was carried to the wet food to be dehydrated via Mylar film. The Mylar film placed on the hot water act as an interface for transfer of thermal heat. In this drying method, the modes of heat transfer: radiation, convection, and conduction. This method facilitates rapid drying due to the process of using heated water below boiling point and a Mylar film with infrared transmission in the wavelength range that matches the water absorption spectrum. A temperature inside the food was usually less than 70°C. The infrared transmission was strongest when a wet product was place on the Mylar film. Due to the optimal combination of product temperature, drying temperature, and drying time, the product obtained by refractance window drying (RWD) shows better retention in quality. This high-quality final product was also partly obtained due to fast mass transfer from the moisture-laden food, resulting in high vapor saturation above the wet food, thereby limiting interaction between product and oxygen. Fruits and vegetables like squash carrots, blueberries, asparagus, mangoes, lingonberries, avocados, strawberries, etc., have been dried using RWD methods [5, 6].

Carrot (*Daucus carota*) was a common root vegetable that was widely used across the world for its high nutritive content. The carrot contains more than 490 phytochemicals and a good amount of vitamin B complex. It contains calcium, potassium, phosphorous, and has high antioxidant activity due to presence of β-carotene. If processed properly, carrot can be converted to the product of high value.

## 8.2   MATERIALS AND METHODS

### 8.2.1   BATCH TYPE REFRACTIVE WINDOW DRYING SET UP

For this project, the patented Refractance Window (RW™) dehydration method was modified from a continuous system to a batch type system. The batch type refractive window drying system was set up by taking a water bath (Lab companion, BW-20G, Figure 8.2) and transparent film which was polyvinyl chloride (Figure 8.3). The water bath has a digital temperature controller with the precision of 0.01°C. The transparent film (polyvinyl chloride) was of area 445 cm², with a thickness of 0.05–0.13 mm and weight about 7–8 g (Figures 8.1–8.3).

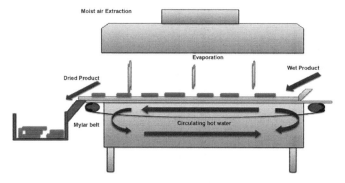

**FIGURE 8.1**   Schematic diagram of continuous refractance window drying system.

**FIGURE 8.2**   Water bath.

**FIGURE 8.3**    Polyvinyl chloride.

### 8.2.2   RAW MATERIAL

Fresh and good quality carrots were procured from the local market situated in Napaam, Tezpur, Assam, India

### 8.2.3   SAMPLE PREPARATION

The carrots procured from the market were peeled, washed with water, and wiped to remove the excess water. The washed carrots were shredded using the vegetable cutter (Crystal, MKA 062, Figure 8.4). The length of the shredded carrots can be varied according to the requirement while the breadth and thickness of the shredded carrot varies, but the range was always between 2.9–3.1 mm and 1.00–1.10 mm, respectively.

**FIGURE 8.4**    Vegetable cutter.

### 8.2.4 PRE-TREATMENT OF CARROT

The pre-treatment involves hot water blanching of shredded carrots. To inactivate peroxidase, the carrots, shredded by using the vegetable cutter, were taken in a muslin cloth and blanched in boiling water (100°C) for 6 min water after which it was cooled down to room temperature using running cold water. The cooled samples were then spread on a sieve to drain of the water [7].

### 8.2.5 DRYING OF SHREDDED CARROTS FOR PRODUCT

Water was filled in the water bath and the temperature was set at 95°C and allowed to heat. The blanched carrots were arranged in the transparent film. When the temperature reaches 95°C the transparent film along with the carrots were kept in the water bath and allowed to be heated for about 75 minutes, after which the dried carrots for the products were obtained (Figures 8.5 and 8.6).

**FIGURE 8.5**   Carrots shreds before drying.

**FIGURE 8.6**   Carrots shreds after drying.

## 8.2.6 DRYING KINETICS

Moisture diffusivity was estimated from the drying curve of the sample being dried. After the water in the water bath reached 95°C, the carrot was placed on the film and left for drying. The samples weight was measured during the uniform interval of time until a constant weight of the sample was reached when the sample attain equilibrium moisture content (EMC). The drying kinetics studies were done by considering the models in Table 8.1. Diffusivity study was done by Fick's second law of diffusion [8].

**TABLE 8.1** Summary of Fitness Models Used for Thin Layer Drying Kinetics

| SL No. | Model | Equation |
|--------|-------|----------|
| 1. | Newton | $MR = exp(-kt)$ |
| 2. | Page | $MR = exp(-kt^n)$ |
| 3. | Henderson and Pabis | $MR = a\ exp(-kt)$ |
| 4. | Logarithmic | $MR = a\ exp(-kt) + c$ |
| 5. | Two Term | $MR = a\ exp(-k_0 t) + b\ exp(-k_1 t)$ |
| 6. | Singh et al. (2014) | $MR = exp(-kt) - akt$ |
| 7. | Approximation of diffusion | $MR = a\ exp(-kt) + (1-a)exp(-kbt)$ |
| 8. | Da Silva et al. (2012) | $MR = (-at - bt^{\frac{1}{2}})$ |
| 9. | Verma et al. (1985) | $MR = a\ exp(-kt) + (1-a)exp(-gbt)$ |
| 10. | Modified Henderson and Pabis | $MR = a\ exp(-kt) + b\ exp(-gt) + c\ exp(-ht)$ |

Fick's second law predicts how diffusion of moisture content from sample causes change in concentration of the sample with time [8]:

$$\frac{\partial \varnothing}{\partial t} = D\frac{\partial^2 \varnothing}{\partial x^2}$$

(1)

where; $\phi$ is concentration (m³), t is time (s), D is the diffusion coefficient (m²/s) and x is length (m).

The moisture equilibrium is constant, when during the drying process, the drying air humidity is constant. Then moisture ratio (MR) is:

$$MR = \frac{M_t - M_e}{M_i - M_e}$$

(2)

where; $M_t$ is moisture content at a particular time, $M_i$ is initial moisture content, and $M_e$ is EMC.

However, when the drying air relative humidity varies continuously, then there is continuous variation in the moisture equilibrium so MR is determined as:

$$MR = \frac{M_t}{M_i} \qquad (3)$$

### 8.2.7  MOISTURE CONTENT ESTIMATION

A known amount of sample was taken and dried in a hot air oven (SELEG TC344, Jain scientific glassworks, Model No: 1458) at $105 \pm 2°C$ for 2 hours to determine the moisture content. Triplicate samples were analyzed to record the average moisture content. The moisture content of the sample was estimated by applying Eqn. (4) [7].

$$\% \text{ Moisture content} = \frac{Weight\,loss}{Initial\,weight} \times 100 \qquad (4)$$

### 8.2.8  CAROTENOID CONTENT ESTIMATION

About 100 mg of dried carrot was taken in a pestle mortar to which 20 ml of acetone was added and pulverized till the sample becomes colorless. The extract was taken in a separating funnel to which 100 ml petroleum ether, a pinch of sodium sulfate was added and shaken for a minute. In the separating funnel, two layers were formed, the upper layer was collected. The lower layer was drained off to a different separating funnel for extraction with petroleum ether. Again, the upper yellow layer was collected. The petroleum ether extract was taken in a volumetric flask and volume made up to 100 ml. The absorbance was measured at 452 nm in a spectrophotometer (Cecil CE7400) to determine β-carotene content of the dried carrot [7]. The amount of carotenoids was calculated by using the following formula [13]:

$$Carotenoids\ content = \frac{A * V\,(ml) * 10^4}{A' * P(g)} \qquad (5)$$

where; A = absorbance, V = total volume extract (ml), A' = 2592 (β-carotene extinction coefficient in petroleum ether) and P = sample weight (g).

### 8.2.9 COLOR MEASUREMENT

The hunter lab calorimeter (Ultra Scan VS) was used for color measurement. In this system, the hunter lab color coordinate system L* (lightness), a* (chromaticity on a green (–) to red (+) axis) and b* (indicates chromaticity on a blue (–) to yellow (+) axis) values was used. Color content was estimated from the results of L, a* and b* values [7].

### 8.2.10 WATER ACTIVITY MEASUREMENT

The water activity of the product was measured by water activity meter (Aqua dew point, 4TE). Samples were placed in the water activity meter at 25°C and the readings were obtained.

### 8.2.11 REHYDRATION RATIO ESTIMATION

5 g of sample was weighed and taken in a 500 ml beaker with 150 ml of distilled water. It was boiled at 100°C for 5 min. After 5 min, the rehydrated sample was weighted. The rehydration ratio was obtained by the following formula [7]:

$$Rehydration\ ratio = \frac{rehydrated\ weight\ of\ the\ dried\ sample}{initial\ weight\ of\ the\ sample} \tag{6}$$

### 8.2.12 TEXTURE ANALYSIS OF THE REHYDRATED CARROTS

After rehydration was done, texture analysis of the rehydrated sample was done in a texture analyzer (Stable microsystem TA HD plus).

### 8.2.13 THE EVALUATION OF THE QUALITY OF CARROT DESSERT PREPARED BY SHREDDED CARROT

Mansur et al. [9] gave a composition of 30 g dehydrated shredded carrot, 20 g milk solid, 40 g sugar, 2.5 g coconut, 5 g dry fruits as having the best quality in terms of cooking time, sensory scores and nutritional values. The same composition was validated for quality parameters. The texture analyzes of the product prepared from fresh and dried carrot were carried out in the texture analyzer (Stable microsystem TA HD plus) and also sensory analyzes were carried out as per the hedonic method (Figure 8.7).

**FIGURE 8.7** Pre-mix.

## 8.3 RESULTS AND DISCUSSION

### 8.3.1 BATCH TYPE REFRACTIVE WINDOW DRYING SET UP

The water bath was filled with water, and the temperature was set at 95°C. After the temperature reaches the required temperature, the temperature of the water bath was measured with a thermometer manually. This was done to check whether the temperature of the water bath will remain constant during the experiment. The temperature of the water bath was then taken at an interval of 30 minutes for 6 hours.

In the batch-type refractive window drying, the temperature of the water bath do not change throughout the time when RWD experiment of shredded carrots were carried out.

The carrots were dried and its moisture content was calculated, and the water activity was measured (Table 8.2).

**TABLE 8.2**   Values Obtain After Drying by Refractance Window Drying Method

| Time (min) | Water Activity (a$_w$) | Moisture Content (%, db) |
|---|---|---|
| 30 | 0.761 | 39.03 |
| 35 | 0.499 | 25.37 |
| 40 | 0.433 | 22.86 |
| 45 | 0.393 | 22.26 |
| 50 | 0.388 | 15.34 |
| 55 | 0.387 | 14.86 |
| 60 | 0.379 | 14.76 |
| 65 | 0.376 | 14.17 |
| 70 | 0.373 | 13.79 |
| 75 | 0.369 | 13.39 |

From an experiment conducted, it has been found that the RWD method was a quick-drying process as compare to tray dryer when dried at 55°C for the same time, i.e., 1 hour and during that period the moisture content was reduced to 89% (dry basis) while it was reduced to less than 23% (dry basis) in RWD method. The time from which the moisture content estimation was done was 30 minutes as during that time, the drying rate was constant, and after 30 minutes, the period of falling drying rate starts. Hence the estimation of water activity and moisture content was started 30 minutes after the beginning of the experiment. The time taken for drying of carrots was 75 min (Figures 8.8 and 8.9).

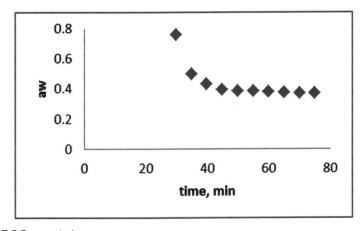

**FIGURE 8.8**   a$_w$ v/s time.

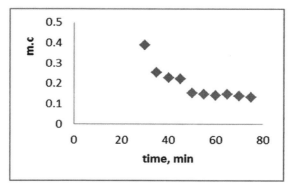

**FIGURE 8.9** M.C v/s time.

### 8.3.2 TEMPERATURE

The temperature was set at 95°C. The temperature of 95°C was use, which gives the appropriate wavelength above 6 μm that gives out high energy for drying the sample. This temperature gives a better-inferred energy which gives better drying.

### 8.3.3 STUDY OF DRYING BEHAVIOR OF SHREDDED CARROTS DRIED BY REFRACTANCE WINDOW DRYING (RWD)

The drying data obtained by the RWD method were fitted to different semi-theoretical models (thin layer drying equations) based on MR for prediction of drying kinetics. The effective diffusivities were obtained from Fick's 2nd law. The thin layer equations were given in Table 8.1. For the study of drying behavior, the data taken was average of triplicate sample (Table 8.3 and Figure 8.10).

**TABLE 8.3** The Values Obtained by Curve Fitting in Thin Layer Models

| Model | Goodness of Fit | | | |
|---|---|---|---|---|
| | SSE | $R^2$ | Adjusted $R^2$ | RMSE |
| Page | 0.022 | 0.977 | 0.976 | 0.037 |
| Midili Kuccuk | 0.021 | 0.978 | 0.974 | 0.039 |
| Weibul | 0.022 | 0.977 | 0.976 | 0.037 |
| Henderson and Pabis | 0.112 | 0.891 | 0.884 | 0.083 |

**TABLE 8.3** *(Continued)*

| Model | Goodness of Fit | | | |
|---|---|---|---|---|
| | SSE | R² | Adjusted R² | RMSE |
| Logarithmic | 0.058 | 0.943 | 0.936 | 0.062 |
| Modified Page | 0.054 | 0.947 | 0.943 | 0.060 |
| Newton | 0.138 | 0.874 | 0.872 | 0.090 |
| Aghabashlo | 0.139 | 0.864 | 0.855 | 0.093 |
| Two-term exponential | 0.096 | 0.906 | 0.900 | 0.077 |
| Verma et al | 0.024 | 0.976 | 0.973 | 0.040 |
| Modified Henderson and Pabis | 0.017 | 0.983 | 0.977 | 0.036 |
| Approximation of Diffusion | 0.024 | 0.976 | 0.973 | 0.040 |

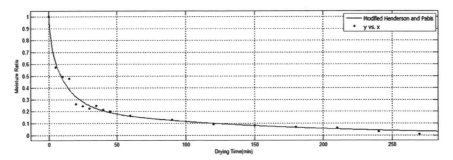

**FIGURE 8.10**   The graph of the best-fit model.

### 8.3.4   *MOISTURE CONTENT ESTIMATION*

The moisture content of fresh carrots and dried shredded carrots were done by the method mentioned in Section 8.2.7 (Table 8.4). The calculations of moisture content were done by Eqn. (4).

**TABLE 8.4**   Moisture Content

| Sample Number | Moisture Content of Raw Carrots (%, wb) | Moisture Content of Shredded Carrots (%, db) |
|---|---|---|
| 1 | 92.23 | 13.15 |
| 2 | 92.55 | 13.19 |
| 3 | 92.02 | 13.25 |
| Average | 92.27 | 13.19 |

The moisture content of raw material was estimated to be 92.26 ± 0.21% (wb) and that of dried carrot was 13.19 ± 0.28% (db). The moisture content obtained by RWD was less as compare to that obtains by tray dryer which gives 55.34 ± 0.02% (db) moisture content when dried at 55C for 75 minutes.

## 8.3.5  CAROTENOIDS CONTENTS

Carotene was an important constituent of carrots. The carotene content of carrots ranges from 600–1200 µg/g. More amount of carotene indicates the carrots to be of good quality. Carrots with more amount of carotene were preferable (Table 8.5). The carotenoids content was calculated by considering the absorbance values and the Eqn. (5).

**TABLE 8.5**   Carotene Content of Dried Carrot

| Sample Number | Carotene Content (µg/g) |
|---|---|
| 1 | 495.29 |
| 2 | 502.85 |
| 3 | 499.07 |
| Average | 499.07 |

The average carotene content was calculated to be 499.07 ± 3.78 µg/g in RWD methods. The carotene content by drying sample in tray dryer at 55°C for 75 minutes was 55 to 70 µg/ g. Therefore, it was concluded that the RWD method gives better retention of carotene content than tray drying method.

## 8.3.6  WATER ACTIVITY

The water activity was measured by the water activity meter. According to scientific data, pathogenic microbes cannot grow at $a_w$ < 0.86; yeasts and molds which were more tolerant usually but no growth occurs at $a_w$ < 0.62 (Table 8.6).

**TABLE 8.6**  Water Activity of Dried Carrot

| Sample Number | Water Activity |
|---|---|
| 1 | 0.3689 |
| 2 | 0.3692 |
| 3 | 0.3690 |
| Average | 0.3690 |

The average water activity was measured to be 0.369 ± 0.0001. This indicates that there was no possible growth of microorganisms as the water activity was below 0.62. In RWD, less water activity was obtained very quickly. From an experiment conducted, it was found that during the first-hour water activity in tray drying product was very high, i.e., 0.65–0.70, while in RWD, it was below 0.62.

### 8.3.7  COLOR MEASUREMENT

The color measurement was done by the hunter lab calorimeter which gives the L, a*, b* values of the sample (Table 8.7).

**TABLE 8.7**  L, a*, b* Values of Dried Carrot

| Sample Number | L | a* | b* |
|---|---|---|---|
| 1 | 59.77 | 5.25 | 1.73 |
| 2 | 59.65 | 5.30 | 1.69 |
| 3 | 59.72 | 5.35 | 1.65 |
| Average | 59.71 | 5.30 | 1.69 |

The average L, a*, b* values of color measurement obtained in RWD was 59.71 ± 0.06, 5.3 ± 0.05 and 1.69 ± 0.04, respectively, while the average values of color measurement obtained by tray drying at 55°C for 75 minutes was 45.76 ± 0.09, 12.45 ± 0.89 and 09.56 ± 0.12, respectively. Hence it was concluded that RWD gives better color retention.

### 8.3.8  REHYDRATION RATIO

The rehydration ratio gives the multiplying factor of by how much the dried carrots will increase when it was rehydrated (Tables 8.8 and 8.9). The ratio was calculated by Eqn. (6).

**TABLE 8.8**  Rehydration Ratio of Dried Carrot by Refractance Window Drying

| Sample Number | Rehydration Ratio |
|---|---|
| 1 | 5.95 |
| 2 | 5.96 |
| 3 | 5.94 |
| Average | 5.95 |

**TABLE 8.9**  Rehydration Ratio of Dried Carrot by Tray Drying

| Sample Number | Rehydration Ratio |
|---|---|
| 1 | 2.25 |
| 2 | 2.30 |
| 3 | 2.35 |
| Average | 2.30 |

The average rehydration ratio was calculated to be $5.95 \pm 0.01$. From experiment conducted, it was found that the rehydration ratio of product dried by RWD was more than that of product obtained from tray drying. The rehydration ratio of product dried at 55°C for 75 minutes in tray dryer was 2.30, which was less than the ratio obtained from the RWD method.

### 8.3.9   TEXTURE PROFILE ANALYSIS OF REHYDRATED CARROTS

The texture profile analysis was done to determine the overall texture of the sample. The texture analysis gives an idea about the hardness, fracturability, adhesiveness, springiness, cohesiveness, gumminess, chewiness, and resilience of the product. From the data obtained, it has been found that during storage, there has been no significant changes in the values obtain during texture profile analysis were given in Table 8.10.

**TABLE 8.10**  Data of Texture Analysis

|  | Sample 1 | Sample 2 | Sample 3 |
|---|---|---|---|
| Hardness | 3586.195 | 3560.178 | 3590.145 |
| Fracturability | 10.60 | 10.7 | 10.5 |
| Adhesiveness | −4.807 | −4.898 | −4.850 |
| Springiness | 0.345 | 0.350 | 0.348 |
| Cohesiveness | 0.455 | 0.454 | 0.454 |
| Gumminess | 1668.982 | 1667.455 | 1667.521 |
| Chewiness | 821.54 | 820.36 | 821.86 |
| Resilience | 0.279 | 0.275 | 0.274 |

### 8.3.10   STUDY OF THE CHANGES IN STORED DRIED CARROTS SHREDS WITH STORAGE TIME

The quality parameters were analyzed. The dried carrots were stored in flexible polythene which was stored for 30 days. The analysis was done at an interval of 15 days. The storage conditions were 75% humidity created by saturated sodium chloride salt at 25°C. During the storage studies, the test for carotene content, water activity, moisture content, measurement of color, rehydration ratio was done by the methods as given in 2.7–2.12.

#### 8.3.10.1   MOISTURE CONTENT

It was found that there was no significant change in the moisture content during storage. The deviation was of the range 0.01% while the moisture content was $13.45 \pm 0.01\%$ (Table 8.11 and Figure 8.11).

**TABLE 8.11**   Moisture Content of Stored Dried Carrot Shreds

| Time (Days) | Day 0 | Day 15 | Day 30 |
|---|---|---|---|
| Moisture content (%, db) | 13.45 | 13.46 | 13.45 |

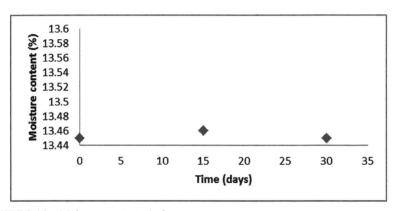

**FIGURE 8.11**   Moisture content v/s time.

#### 8.3.10.2   CAROTENOIDS CONTENT

The average of carotene content during the storage time was calculated to be $501.18 \pm 1.9$ µg/g. The deviation in carotene content was estimated to be $\pm 1.9$ µg/g in 30 days (Table 8.12 and Figure 8.12).

**TABLE 8.12** Carotene Content of Stored Dried Carrots

| Time (Days) | Day 0 | Day 15 | Day 30 |
|---|---|---|---|
| Absorbance | 0.133 | 0.132 | 0.134 |
| Carotene content (µg/g) | 501.85 | 499.07 | 501.63 |

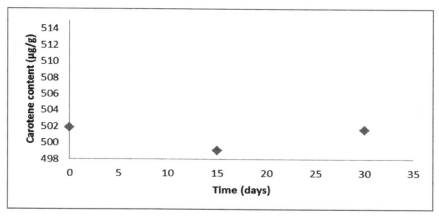

**FIGURE 8.12** Carotene content v/s time.

### 8.3.10.3 COLOR MEASUREMENT

The color measurement was done for each sample at an interval of 15 days. The results obtained showed no significant change in color (Table 8.13 and Figure 8.13).

**TABLE 8.13** Color Measurement of Stored Dried Carrots

| Time (Days) | L | a* | b* |
|---|---|---|---|
| Day 0 | 59.771 | 5.255 | 1.733 |
| Day 15 | 59.772 | 5.255 | 1.732 |
| Day 30 | 59.771 | 5.256 | 1.733 |

### 8.3.10.4 REHYDRATION RATIO

There were no significant changes observed in the rehydration ratio during the storage study of rehydration ratio of the dried shredded carrots (Table 8.14 and Figure 8.14).

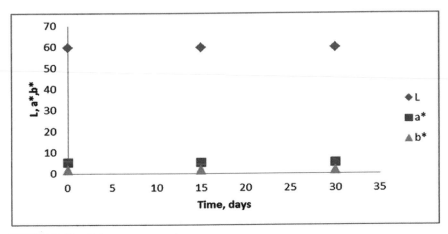

**FIGURE 8.13**    Color measurement.

**TABLE 8.14**    Rehydration Ratio of Stored Dried Carrots

| Time (days) | Day 0 | Day 15 | Day 30 |
|---|---|---|---|
| Rehydration ratio | 5.98 | 5.99 | 5.99 |

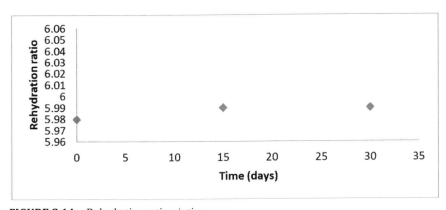

**FIGURE 8.14**    Rehydration ratio v/s time.

## 8.3.10.5    WATER ACTIVITY

There were no significant changes observed in the water activity of stored dried carrots during the storage study of dried shredded carrots (Table 8.15 and Figure 8.15).

**TABLE 8.15**   Water Activity of Stored Dried Carrots

| Time (Days) | Day 0 | Day 15 | Day 30 |
|---|---|---|---|
| Water activity | 0.368 | 0.368 | 0.369 |

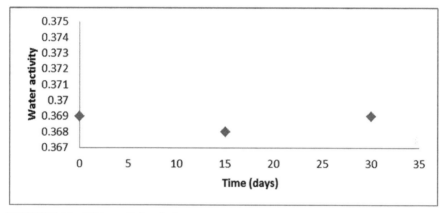

**FIGURE 8.15**   Water activity v/s time.

## 8.3.11   *EVALUATION OF QUALITY OF CARROT DESSERT PREPARED BY SHREDDED CARROTS*

The pre-mix was created with the combination of dried shredded carrots, milk solids, and sugar. The pre-mix was cooked in a pressure cooker, and the dessert was prepared. The dessert was then analyzed for texture and sensory evaluation.

### 8.3.11.1   *TEXTURE PROFILE ANALYSIS OF PRE-MIX CARROTS DESSERTS*

The texture profile analysis was done to determine the overall texture of the sample (Figure 8.16). The texture analysis gave an idea about the hardness, fracturability, adhesiveness, springiness, cohesiveness, gumminess, chewiness, and resilience of the product. From the data obtained, it has been found that during storage, there has been no significant changes in the values obtain during texture profile analysis were given in Table 8.16.

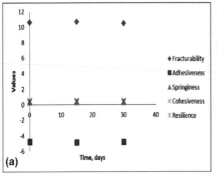
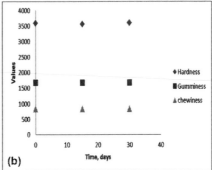

**FIGURE 8.16**   Texture profile analysis (a & b).

**TABLE 8.16**   Data of Texture Analysis

| Time (Days)   | Day 0     | Day 15    | Day 30    |
|---------------|-----------|-----------|-----------|
| Hardness      | 3586.195  | 3560.178  | 3590.145  |
| Fracturability| 10.60     | 10.7      | 10.5      |
| Adhesiveness  | −4.807    | −4.898    | −4.850    |
| Springiness   | 0.345     | 0.350     | 0.348     |
| Cohesiveness  | 0.455     | 0.454     | 0.454     |
| Gumminess     | 1668.982  | 1667.455  | 1667.521  |
| Chewiness     | 821.54    | 820.36    | 821.86    |
| Resilience    | 0.279     | 0.275     | 0.274     |

### 8.3.11.2   SENSORY ANALYSIS OF PRE-MIX CARROTS DESSERTS

Sensory analysis was done as per the hedonic scale. Here the average values of a wider survey were taken into consideration.

From the result of the sensory analysis, it can be concluded that the product, i.e., the carrot dessert prepared from the pre-mix was of good quality and the acceptability of the product were positive (Table 8.17).

**TABLE 8.17**   Sensory Analysis

|                       | Day 0 | Day 15 | Day 30 |
|-----------------------|-------|--------|--------|
| Color                 | 9     | 8      | 9      |
| Texture               | 8     | 8      | 8      |
| Flavor                | 8     | 9      | 7      |
| Overall acceptability | 8     | 9      | 8      |

## 8.4    CONCLUSIONS

The project aimed at producing dried shredded carrot by RWD method and characterizing the drying behavior as well as the dried product. The carrots were shredded with a cutter to a size of about 1.00–1.10 mm, dried under a batch type laboratory set up of RWD system. Moisture loss behavior was modeled to one of the empirical models based on statistics of the fit. The enhancement in retention of carotene was verified by chemical measurement of carotene content of samples dried with the system.

From the results of the experimentation of the project, the following conclusions were made:

1.  Refractance window drying method was a faster method of drying in comparison to conventional drying; moisture content of shredded carrot could be reduced to less than 13% (db) in 75 mins.
2.  Water activity achieved after one hour of drying was less than 0.62.
3.  The retention of carotene content was better. By this method carotene content retained was found to be 499.07 ± 3.78 µg/g while in that of tray-dried sample was 55 to 70 µg/ g.
4.  Analysis of the moisture loss kinetics during the refractance window drying of shredded carrot shows that Modified Henderson and Pabis model fitted better, with $R^2$ at 0.983 and RMSE at 0.03673.
5.  The rehydration ratio of the product dried by refractive window drying method was better as it was in the range of 5.95 ± 0.01.
6.  The texture analysis of the rehydrated carrot gave good and acceptable result.

## KEYWORDS

- carotene content
- equilibrium moisture content
- Fick's second law
- moisture ratio
- refractance window drying method
- rehydration ratio

## REFERENCES

1. Abbasid, A., Niakousari, M., & Yasini, A. S. A., (2015). The advantages of the refractance window method of dehydrating fresh tomato slices and the relevant characteristics thereof. *Journal of Applied Environmental and Biological Sciences*, *4*, 6–13.

2. Ochoa-Martínez, C. I., Quintero, P. T., Ayala, A. A., & Ortiz, M. J., (2012). Drying characteristics of mango slices using the refractance window™ technique. *Journal of Food Engineering*, *109*(1), 69–75.

3. Akinola, A. A., Lawal, S. O., & Osiberu, A. S., (2014). Refractance window™ drying of red onions (*Allium cepa*).

4. Ocoro-Zamora, M. U., & Ayala-Aponte, A. A., (2013). Influence of thickness on the drying of papaya puree (*Carica papaya* L.) through refractance window technology. *Dyna*, *80*(182), 147–154.

5. Nindo, C. I., Feng, H., Shen, G. Q., Tang, J., & Kang, D. H., (2003). Energy utilization and microbial reduction in a new film drying system, *Journal of Food Processing Preservation 27*, 117–136.

6. Nindo, C., Sun, T., Wang, S. W., Tang, J., & Powers, J. R., (2003). Evaluation of drying technologies for retention of physical quality and antioxidants in asparagus (*Asparagus officinalis*, L.). *LWT-Food Science and Technology*, *36*(5), 507–516.

7. Prakash, S., Jha, S. K., & Datta, N., (2004). Performance evaluation of blanched carrots dried by three different driers. *Journal of Food Engineering*, *62*(3), 305–313.

8. Goyal, R. K., Kingsly, A. R. P., Manikantan, M. R., & Ilyas, S. M., (2007). Mathematical modeling of thin-layer drying kinetics of plum in a tunnel dryer. *Journal of Food Engineering*, *79*(1), 176–180.

9. Mansoor, G. S., Khursheed, A., & Jairajpuri, D., (2013). Preparation, processing, and packaging of pre-mix for the production of carrot dessert. *Journal of Environmental Science, Toxicology, and Food Technology*, *3*(6).

10. Singh, F., Katiyar, V. K., & Singh, B. P., (2014). *Mathematical Modelling to Study Drying Characteristic of Apple and Potato* (pp. 172–175). International Conference on Chemical, Environment and Biological Sciences, Kuala Lumpur.

11. Da Silva, W. P., Silva, C. M. D. P. S., Farias, V. S. O., & Gomes, J. P., (2012). Diffusion models to describe the drying process of peeled bananas: Optimization and simulation. *Drying Technology, 30*, 164–174. https://doi.org/10.1080/07373937.2011.628554.

12. Verma, L. R., Bucklin, R. A., Endan, J. B., & Wratten, F. T., (1985). Effects of drying air parameters on rice drying models. *Transactions of the ASAE, 28*, 296–301. https://doi.org/10.13031/2013.32245.

13. De Carvalho, L. M. J., Gomes, P. B., De Oliveira Godoy, R. L., Pacheco, S., Do Monte, P. H. F., De Carvalho, J. L. V., & Ramos, S. R. R., (2012). Total carotenoid content, α-carotene and β-carotene, of landrace pumpkins (*Cucurbita moschata Duch*): A preliminary study. *Food Research International, 47*(2), 337–340.

# CHAPTER 9

# Valorization of Biomass from Tea Processing

ERICK M. PEÑA-LUCIO,[1] MÓNICA L. CHAVEZ-GONZALEZ,[1]
LILIANA LONDOÑO-HERNANDEZ,[2] JOSÉ L. MARTÍNEZ-HERNANDEZ,[1]
MAYELA GOVEA-SALAS,[1] HECTOR RUIZ-LEZA,[1]
ABDULHAMEED SABU,[3] and CRISTÓBAL N. AGUILAR[1]

[1]*Bioprocesses and Bioproducts Research Group, Food Research Department, Autonomous University of Coahuila, Saltillo – 25280, Mexico, E-mail: cristobal.aguilar@uadec.edu.mx (C. N. Aguilar)*

[2]*Applied Microbiology and Biotechnology Research Group, Department of Biology, Universidad del Valle, Cali, Colombia*

[3]*Department of Biotechnology and Microbiology, Kannur University, Thalassery Campus, Kannur – 670661, India*

## ABSTRACT

Tea is amply consumed worldwide and, depending on its degree of fermentation, can be classified in green, black, or oolong. This natural beverage has a wide variety of properties due to its bioactive components; however, waste produced can impact on the environment. In recent years, the leaves of tea spent have been used with the aim of reducing its ecological impact. Its high content of polyphenols gives it excellent properties to be used as substrate to the application of bioprocesses with the aim of obtaining secondary metabolites; however, one of the most attractive alternatives in the production of new sources of energy such as biogas, which can be obtained in an economical and simple way.

## 9.1   INTRODUCTION

The tea is originated from China, and it can be obtained from the plant *Camellia sinensis*. This term was coined in 2737 BCE by Shen Nung for the first time that it was made, some leaves of tea were steamed [1]. This plant is a shrub or small perennial tree that has green leaves of 4 to 15 cm in length and yellowish-white flowers [1]. The tea (*Camelia sinensis*) is extensively consumed worldwide for its health benefits. The origin of the green tea goes back to China in 2008, and it was responsible for around 73% of the world's production; however, it is currently produced in about 20 countries whose tropical and subtropical climates allow their cultivation [1]. Currently, the annual World production of green tea has increased by around 7.5%. It was promoted mainly by China, for 2023 is expected production of 2,970,000 tons, displaying a range in the production of 3.6 million of tons for 2027 [2].

However, the percentage of soluble compounds (polyphenols) differs according to the type of tea; for example, the black tea contains less percentage of polyphenols than green tea [3].

Black tea is a kind of tea that comes from *Camellia sinensis*, neither green nor white tea are completely fermented. However, black tea has been classified as a tea fermented totally, because, the flavonoids present in these compounds are fermented and condensed via enzymatic [4].

Depending on the geographical location have classified some varieties of black tea and currently its consumption reaches about 78%. The white tea is originated from the *Camellia sinensis* plant, it reaches a very low fermentation level and is produced using only plant sprouts, three subtypes can be described; silver needles, white peony and Show Mei, each with different chemical and biological characteristics [5], there is other kind of green tea that is realized in autumn; however, it has some undesirable properties, it is due to astringent taste and low aroma, in consequence, it is produced in low quantities [5].

## 9.2   CHEMICAL COMPOSITION

The chemical content of tea depends on the plant variety and its processing; It has been reported that green tea contains some quantities of polyphenols (32.40 mg/mL), caffeine (11.28 mg/mL), free amino acids (0.69 mg/mL), epigallocatechin gallate (EGCG) (16.07 mg/mL) and epigallocatechin (7.67 mg/m) [5]. Leaves of green tea contain antioxidative catechins (epicatechin (EC), epigallocatechin (EGC), epicatechin gallate (ECG)) and the most abundant catechin present in the green tea is ECG [5, 6], which forms around 50% of catechin content in tea [7], in comparison, the tea byproduct has a great content of tannins, carbohydrates, i.e., cellulose, and hemicellulose, in consequence, agro-industrial residues from the green tea can be approached with the aim to use it as support to fermentation [8, 10]. Black tea has five different types of catechins (Table 9.1), each with different concentrations (+ C). It is found in concentrations around 18 μmol/L, EC 27 μmol/L, ECG 14 μmol/L, EGC 28 μmol/L, EGCG 20 μmol/L [6]. More than 99 compounds have been found in the white tea, within which 5 alkaloids can be identified, 12 amino acids, 12 catechins, 9 nucleosides, 11 phenolic acids, 5 aroma precursors, 13 dimers of catechin, and 29 glycosides of flavonols. More than 60% of the polyphenols that compose tea are catechins, within which you can find Proanthocyanidins (PAs) (Procyanidin B1 and Procyanidin B2) and Theasinensins [9].

**TABLE 9.1**  Different Polyphenol Compounds in Different Kinds of Tea

| Type of Tea | Type of Polyphenol | References |
|---|---|---|
| Green tea | Kaempferol, quercetin, myricetin, rutin, nicotiflorine, gallic acid, (–)-epigallocatechin 3-O-gallate, myricetin 3-O-β-D-galactopyranoside, myricetin 3-O-β-D-glucopyranoside, (–)-gallocatechin 3-O- gallate, (–)-epicatechin 3-O-gallate, (–)-catechin 3-O-gallate, isovitexin, 6″-galloylmyricetin 3-O-β-D-glucopyranoside, quercetin 3-O-β-D-galactopyranoside, vitexin, quercetin 3-O-β-D- glucopyranoside, (-)-epicatechin, myricetin 3-O-rutinoside, 6″-galloylmyricetin 3-O-β-D-galactopyranoside, daidzin, tricin 7-O-β-D-glucopyranoside, quercetin 3-O-[β-D-glucopyranosyl- (1 → 4)-O-α-L-rhamnopyranosyl-(1 → 6)-O-β-D-glucopyranoside], quercetin 3-O-[β-D- glucopyranosyl-(1 → 4)-O-α-L-rhamnopyranosyl-(1 → 6)-O-β-D-galactopyranoside], kaempferol 3-O-[β-D-glucopyranosyl-(1 → 3)-O-α-L-rhamnopyranosyl-(1 → 6)-O-β-D-glucopyranoside], kaempferol 3-O-[β-D-glucopyranosyl-(1 → 3)-O-α-L-rhamnopyranosyl-(1 → 6)-O-β-D- galactopyranoside], luteolin 8-C-glucopyranoside, quercetin 3-O-[2-O″-(E)-p-coumaroyl][β-D-glucopyranosyl-(1 → 3)-O-α-L-rhamnopyranosyl-(1 → 6)-O-β-D-glucopyranoside] | [9, 11, 15] |

**TABLE 9.1**   *(Continued)*

| Type of Tea | Type of Polyphenol | References |
|---|---|---|
| White tea | flavonol-3-O-glycosides, glucoside galgalactoside, myricetin-3-O-rhamnodiglucoside, myricetin-3-O-glucorhamno-glucoside, myricetin-3-O-rutinoside, M-galmyricetin-3-O-galactoside, myricetin-3-O-glucoside, myricetin-3-O-diglucorhamnoglucoside, quercetin-3-O-galactorhamnoglucoside, rutquercetin-3-O-glucorhamnoglucoside, quercetin-3-O-rutinoside, quercetin-3-O-galactoside, quercetin-3-O-glucoside, kaempferol-3-O-galactorhamnoglucoside, kaempferol-3-O-glucorhamnoglucoside, kaempferol-3-O-galactoside, kaempferol-3-O-rutinoside, kaempferol-3-O-glucoside, epigallocatechin, epicatechin, epigallocatechingallate, epicatechin gallate, gallocatechin, catechin | [12, 13] |
| Black tea | (−)-gallocatechin; (−)-epigallocatechin; (+)-catechin; (−)-epigallocatechin gallate; (−)-epicatechin; (−)-gallocatechin gallate; (−)-epicatechin gallate. | [14] |

## 9.3   BIOLOGICAL PROPERTIES OF TEA

In addition to its traditional use as a beverage, today, tea is also used to obtain bioactive compounds. It has some estates, such as antibacterial, anticarcinogenic, antimutagenic, anticarcinogenic, antioxidant, antiarthritic, anticarcinogenic, antiviral, antifungal, and anticoccidial activity [6, 15], Some of the compounds found in tea leaves (fermented and unfermented) are phenolic compounds such as catechins, which are recognized for their biological antioxidant, anti-cancer, cardiovascular protector, anti-hypercholesterolemia, anti-obesity activities [9, 16]. For example, EGCG prevents cancer and improves colon carcinogenesis in rats [16]. Likewise, it has been found that extracts from tea leaves are rich in policosanol [15].

Green tea consumption can avoid cardiovascular diseases and decreased serum total cholesterol [17]. Its phenolic compounds become the major agents to present antioxidant activity, and it is due to their act as reducing agents [5]. The antioxidant activity is better when it has a major number of hydroxyl groups present [17]. In addition, green tea extract can serve to reduce the aging process and hypercholesterolemia; furthermore, catechins repress tyrosinase, collagenase, and elastase activities, it promoting skin health. White tea presents numerous benefits, such as antimutagenic, anticancerogenic, and antibacterial activity. Theasinensis is a type of catechin presents in the white tea that has an interesting pharmacological effect; it has presented antioxidative effects [17]. Black tea has the ability to decrease the levels of Nε-(carboxymethyl) lysine (CML) and Nε-(carboxyethyl) lysine (CEL), these compounds are in cereals, meats, and nuts [17]. They have been related with the pathogenesis of arteriosclerosis, diabetes mellitus, uremia, heart failure, and neurodegenerative disorders [13]. The presence of polyphenols in black tea can inhibit autoxidation in some compounds. As a consequence, it can avoid some pathologies controlling cholesterol and increasing HDL cholesterol levels [17].

## 9.4  COMMERCIAL PRODUCTS OBTAINED FROM TEA

Tea (*Camellia sinensis*) is an evergreen tree or shrub originating in China [18]. Commercially, for many years tea leaves have been used to obtain a traditional drink, which is recognized worldwide for its functional properties and its positive effects on consumer health [18]. According to the processing, four types of tea are recognized: green tea is the drink made from unfermented leaves, the oolong tea in which the leaves are partially fermented, and black and red tea, in which the leaves used have been subject to a fermentation process [19]. These compounds have been extracted through different technologies and are being incorporated into food matrices to obtain *functional foods* or the manufacture of *nutraceuticals* [19]. Tealeaf extract has been used in some food products such as biscuits, bread, dried apples, and meat products, finding, among other benefits, that it increases the shelf life of the product by increasing stability and preventing oxidation of fats [20]. Taking advantage of these antioxidant benefits of phenolic compounds [21-23], the reported properties of protection against UV radiation, and the decrease in the risk of skin cancer, *sunscreens*, and *skin creams* have been developed using tea leaf extracts [24–26]. In addition to these commercial uses of tea leaf extracts, tea seed kernels are being used for obtaining *oils* [27]. These oils are composed of unsaturated fatty acids, and they are considered bioactive compounds [27]. These oils are being included in food matrices, creams or cosmetic products, and the obtaining of pharmaceuticals [27].

## 9.5  USES OF HOT, COLD, OR FREEZE-DRIED TEA BIOMASS

In recent years with the increase in the cultivation and consumption of tea leaves, by-products from hot, cold or freeze-dried processing have been increasing [27]. These residues are not used for a specific purpose and are left in the area, which can cause environmental problems if they decompose. Some reports indicate that tea residues can be used for the production of bioethanol and other biofuels [28]. In this sense, tea residues have also been used to obtain hydrogen and other gaseous compounds of interest [28, 29]. Another application of growing interest in the processing of tea biomass is the obtaining of *microcrystalline cellulose*, which has various applications in the food industry (anti-caking agent, texture agent, foam control agent) [30].

## 9.6  AGROINDUSTRIAL RESIDUES FROM TEA INDUSTRY

Agro-industrial waste can generate some environmental problems, an alternative to diminish its ecological impact is its bioconversion through bioprocess, such as, solid state fermentation [30].

The excessive cultivation of Summer and Autumn tea plants generates millions of tons of waste [13]. However, these have similar levels of Bioactive compounds; in consequence, from this waste can be generated theaflavins [15], which are present in different functions pharmacological, and it can be used as a treatment against various cardiovascular conditions, it also presents antimicrobial and anticancerigen activity [15].

India ranks Second in the world tea production, and annually are consumed 3.5 million of tea [2], as a result, daily are generated large amounts of waste that are deposited in containers, currently has been studied its application and it has been reported a high content of tannins, it which can be converted to galic acid [15], another application that can be given to the waste is the bioremediation from heavy metals [13].

Tea waste has been used to enzymatic production, obtaining good yields, because this material contains polyphenols and tannins that induce tannase production [34], although, other types of substrates can be used, it has been reported an increase in the yield of up to 18.9% by applying leaves of tea spent by means of solid state fermentation [31]. Similarly, it has also been produced tannase using *Aspergillus* sp., low fermentation in solid-state with green tea [32].

During production there may be some incidents, due to this, not all the plants that are sown can be harvested, generating losses and waste, traditionally the agroindustrial residues serve as fertilizers for the land, however, it also can be applied to the generation of biofuels [29]. The energies that form carbon sources are going through a radical decrease, coupled with this, they can impact negatively on the environment, the production of energy from other sources can be a way to reduce pollution and generate energy alternatives [29]; It is possible to produce biogas by means of the pyrolysis of the waste of the tea, for its obtaining there are some procedures, such as catalytic and noncatalytic ways [28]; It has been reported that is possible to achieve yields near 57.49%, this can vary, depending on some factors like the concentration, temperature, and pH, in some process the optimization of the generation of biogas has been studied obtaining that a temperature around 850°C has favored the formation of biogas [28].

Currently, materials that can be used as absorbents to remove contaminants have been investigated, for example, industrial byproducts, seaweed residues, alumina, clay, cob, a low cost alternative and easy maintenance are spent tea leaves [13], which can be recovered after its processing, although, these residues have to go through an activation process, the removal of water contaminants using residues of tea leafs can be applied without any type of processing [28].

## 9.7   BIOPROCESSES DEVELOPED BASED ON TEA WASTE

There are different types of bioprocesses used for bioremediation, biolixiviation, and bio beneficiation; within these are solid and liquid fermentation [32]. This first can be defined as a biological process that occurs on a solid support with the absence of water disponible and in aseptic and natural conditions, the second one uses aqueous media and has the advantage that some conditions such as pH, agitation, and temperature are easily controlled [33]. Most of the bioprocesses performed require the biotechnological use of microorganisms that have the ability to take advantage of the substrates present to develop and reproduce [33]; there is a wide variety of fungal and bacterial microorganisms that have been used in fermentation in solid and submerged state, for example, *Aspergillus* sp., *Trichoderma* sp., *Bacillus* sp., *Lactobacillus* sp., and *Streptococcus* sp. [34].

In recent years, agro-industrial waste has been used as a matrix for the formation of products with high added value, either for animal feed or for obtaining enzymes of industrial interest, at low cost [30]. There is a great variety of residues that can be used as support, from materials with high content of carbohydrates, proteins, and lipids to substrates with high percentages of polyphenols [31]. The green tea residue has been reported as a material with a high content of tannic acid [31], in addition, 75–80% of the polyphenols contained in the green tea are catechins, including epicatechin, epigallocatechin, ECG and EGCG [31], the hydrolysis and biotransformation of these compounds have some advantages because is possible to synthesize some important compounds (such as gallic acid), in addition, some secondary metabolites can be obtained, such as, tannin acyl hydrolase [34], one of its most important characteristics is that it has shown an interesting activity on complex polyphenols, such as, ECG, EGCG, chlorogenic acid, etc. Furthermore, tannase activity has been studied for its capability to improve the chelation of metals ions and the antioxidant activity [35].

Tannase (EC 3.1.1.20) catalyze the bioconversion of ester bonds present in complex tannins releasing other compounds [34]. Some fungus can degradate tannins, the degradation of gallotanins has been studied in *Aspergillus sp;* for example, in *Aspergillus niger*, gallic acid is converted to tricarboxylic acid and then it forms citric acid [34].

## 9.8   PRODUCTION OF FUELS THROUGH THE BIOMASS OF THE TEA

Today, energy sources worldwide come from fossils fuels; it is estimated that daily more than 5 trillion gallons per day is required; however, the growing deterioration of the environment, the increase in gas emissions, and the Depletion of fossils sources, in consequence, is important to find other forms of fuel production [36]. An attractive alternative for energy generation is the use of biomass from different materials, such as rice, jute, coconut, cotton, sugarcane, coffee pulp, and tea, as sources to produce biofuels. In addition, it is worth mentioning that the production of waste has increased; consequently, its use confers its added value and decreases environmental pollution [36] Organic matter residues are present in different forms; they can be found as biomass residues, leaves, seeds, branches, and logs [37]. Annually are produced around 800,000 tons of tea, both the tea industry and the organic wastes that are wasted in houses, hotels, and restaurants are an interesting substrate to solid-state fermentation and also for the production of biofuels [37].

The production of biogas has been reported using the leaves of tea spent, this process is efficient and economic, in addition to reducing the emissions of gases to the atmosphere [29]. Spent tea waste has been used to produce methane gas ($CH_4$) by means of anaerobic fermentation; some studies have shown different yields of production, depending on some factors such as the influence of pH, the radius of carbon nitrogen [29]. The pH values may vary throughout the anaerobic fermentation, the highest yields of biogas generation have been reached at pH values between 7 and 8.5 [32]. However, the hydrolysis of organic matter can generate a decrease of pH influencing significantly in the process [32]. The production of biogas requires a relationship of C/N ratio of around 20:30, because lower levels can affect the bacterial growth and cause accumulation of ammonium NH3, in consequence, poisoning in bacteria [36].

## ACKNOWLEDGMENTS

Authors thank the financial support of fund FONCICYT-CONACYT-Mexico-India through project No. 266614.

## KEYWORDS

- **biocombustible**
- **biotransformation**
- **epicatechin gallate**
- **low-density lipoprotein**
- **phytochemical**
- **tea biomass**

## REFERENCES

1. Saeed, M., et al., (2017). Green tea (*Camellia sinensis*) and L-theanine: Medicinal values and beneficial applications in humans: A comprehensive review. *Biomedicine and Pharmacotherapy*, 1260–1275.
2. FAO, (2019). FAO + China. *Partnering for Sustainable Food Security*. Retrieved from: http://www.fao.org/3/ca4948en/ca4948en.pdf (accessed on 13 January 2021).
3. Sánchez, E. P., et al., (2013). *Hepatoxicity Due to Green Tea Consumption (Camellia sinensis): A Review* (pp. 2, 3). Colombian gastroenterology associations.
4. Zhang, Y. N., et al., (2016). Improving the sweet aftertaste of green tea infusion with tannase. *Food Chemistry*, 470–476. http://dx.doi.org/10.1016/j.foodchem.2015.07.046.
5. Zheng, Y., Xu, X., & Zou, X., (2016). Biotransformation of caffeine in oolong tea by *Paecilomyces gunnii*. *International Biodeterioration & Biodegradation, 114*, 141–144. https://doi.org/10.1016/j.ibiod.2016.04.013.

6. Firmani, P., Silvia, D. L., Remo, B., & Federico, M., (2019). Near-infrared (NIR) spectroscopy-based classification for the authentication of Darjeeling black tea. *Food Control.*, (1), 123–129.

7. Battestin, V., Macedo, G. A., & Freitas, V. A. P. D., (2008). Hydrolysis of epigallocatechin gallate using a tannase from *Paecilomyces variotii. Food Chemistry,* (1), 228–233.

8. Bhushan, B., Mahato, D. K., Verma, D. K., Kapri, M., & Srivastav, P. P., (2018). Potential health benefits of tea polyphenols: A review. In: Goyal, M. R., & Verma, D. K., (eds.), *Engineering Interventions in Agricultural Processing: As Part of Book* series on *Innovations in Agricultural and Biological Engineering* (Vol. 8, pp. 229–282). Apple Academic Press, USA.

9. Tan, J., et al., (2017). Flavonoids, phenolic acids, alkaloids, and theanine in different types of authentic Chinese white tea samples. *Journal of Food Composition and Analysis*, 8–15. Academic Press. doi: 10.1016/J.JFCA.2016.12.011.

10. Joo, H. B., Hyung, J. S., So, Y. C., Yooheon, P., & Hyeon-Son, C., (2014). Differential activities of fungi-derived tannases on biotransformation. *Cellular Biochemistry*, 3, 4.

11. Ni, H., et al., (2015). Biotransformation of tea catechins using *Aspergillus niger* tannase prepared by solid-state fermentation on tea byproduct. *LWT-Food Science and Technology*, 1206–1213. http://dx.doi.org/10.1016/j.lwt.2014.09.010.

12. Hong, Y. D., et al., (2019). Identification of fermented tea (*Camellia sinensis*) polyphenols and their inhibitory activities against amyloid-beta aggregation. *Phytochemistry*, 11–18. Elsevier. doi: 10.1016/j.phytochem.2018.12.013.

13. Parameswaran, B., Varjani, S., & Raveendran, S., (2015). *Green Bio-Processes.* https:// doi.org/https://doi.org/10.1007/978-981-13-3263-0.

14. Azevedo, R. S. A., et al., (2019). Multivariate analysis of the composition of bioactive in tea of the species Camellia sinensis. *Food Chemistry,* 39–44. Elsevier.

15. Lorenzo, J. M., & Munekata, P. E. S., (2016). Phenolic compounds of green tea: Health benefits and technological application in food. *Asian Pacific Journal of Tropical Biomedicine,* 709–719.

16. Lima, J. S., Cabrera, M. P., De Souza Motta, C. M., Converti, A., & Carvalho, L. B., (2018). Hydrolysis of tannins by tannase immobilized onto magnetic diatomaceous earth nanoparticles coated with polyaniline. *Food Research International, 107*(2017), 470–476. https://doi.org/10.1016/j.foodres.2018.02.06.

17. Sun, L., et al., (2019). Comparative effect of black, green, oolong, and white tea intake on body weight gain and bile acid metabolism. *Nutrition*, 12–14. Elsevier. doi: 10.1016/J.NUT.2019.02.006

18. Singh, V., Verma, D. K., & Mahato, D. K., (2018). Biochemical composition, processing technology, and health benefits of green tea: A review. In: Verma, D. K., & Goyal, M. R., (eds.), *Engineering Interventions in Foods and Plants: As Part of Book Series on Innovations in Agricultural and Biological Engineering* (pp. 119–156). Apple Academic Press, USA.

19. Clement, Y., (2009). Can green tea do that? A literature review of the clinical evidence. *Preventive Medicine, 2/3*, 83–87.

20. Choi, S. J., et al., (2016). Contents and compositions of policosanols in green Tea (*Camellia sinensis*) leaves. *Food Chemistry*, 94–101.

21. Sharangi, A. B., (2009). Medicinal and therapeutic potentialities of tea (*Camellia sinensis* L.): A review. *Food Research International*, 529–535.

22. Jain, A., et al., (2013). Tea and human health: The dark shadows. *Toxicology Letters,* 82–87.

23. Lavelli, V., Claudia, V., Mark, C., & William, K., (2010). Formulation of a dry green tea-apple product: Study on antioxidant and color stability. *Journal of Food Science, 75*(2), 184–190.

24. Mitsumoto, M., O'Grady, M. N., Joe, P. K., & Joe, B. D., (2005). Addition of tea catechins and vitamin C on sensory evaluation, color and lipid stability during chilled storage in cooked or raw beef and chicken patties. *Meat Science,* 773–779.

25. Namal, S. S. P. J., (2013). Green tea extract: Chemistry, antioxidant properties and food applications: A review. *Journal of Functional Foods,* 1529–1541.

26. Neethu, R. S., & Pradeep, N. S., (2018). *Isolation and Characterization of Potential Tannase Producing Fungi from Mangroves and Tanneries, 21*(3), 1–13.

27. Wang, R., & Weibiao, Z., (2004). Stability of tea catechins in the bread-making process. *Journal of Agricultural and Food Chemistry,* 8224–8229.

28. Ayas, N., & Tugce, E., (2016). Hydrogen production from tea waste. *International Journal of Hydrogen Energy, 19,* 72.

29. Pütün, A. E., et al., (2009). Synthetic fuel production from tea waste: Characterization of bio-oil and bio-char. *Fuel Combustibles,* 176–184.

30. Saval, S., et al., (2012). Use of agro-industrial waste, past, present and future. *BioTechnology,* 14–46.

31. Noh, D. O., Hyeon, S. C., & Hyung, J. S., (2014). Catechine biotransformation by tannase with sequential addition of substrate. *Process Biochemistry,* 271–276. http://dx.doi.org/10.1016/j.procbio.2013.11.001.

32. Chávez-González, M. L., et al., (2014). Production profiles of phenolics from fungal tannic acid biodegradation in submerged and solid-state fermentation. *Process Biochemistry,* 541–546. http://dx.doi.org/10.1016/j.procbio.2014.01.031.

33. Thomas, L., Larroche, C., & Pandey, A., (2013). Current developments in solid-state fermentation. *Biochemical Engineering Journal, 81,* 146–161. https://doi.org/10.1016/j.bej.2013.10.013.

34. Chávez-González, M., et al., (2012). Biotechnological advances and challenges of tannase: An overview. *Food and Bioprocess Technology, 2,* 445–459.

35. Helak, B., et al., (2019). Antioxidative potential, nutritional value and sensory profiles of confectionery fortified with green and yellow tea leaves (*Camellia sinensis*). *Food Chemistry,* 448–454.

36. Dahiya, S., Kumar, A. N., Sravan, J. S., Chatterjee, S., Sarkar, O., & Mohan, S. V., (2018). Bioresource technology food waste biorefinery: Sustainable strategy for circular bioeconomy. *Bioresource Technology, 248*(2017), 2–12. https://doi.org/10.1016/j.biortech.2017.07.176.

37. Jianguo, J., Changxiu, G., Jiaming, W., Sicong, T., & Yujing, Z., (2014). Effects of ultrasound pre-treatment on the amount of dissolved organic matter extracted from food waste. *Bioresour. Technol., 155,* 266–271.

# CHAPTER 10

# Maize Gelling Arabinoxylans Isolated by a Semi-Pilot Scale Procedure: Viscoelastic Properties and Microstructural Characteristics

JOSÉ MIGUEL FIERRO-ISLAS,[1] MARCEL MARTÍNEZ-PORCHAS,[1] RAFAEL CANETT ROMERO,[2] FRANCISCO BROWN-BOJORQUEZ,[2] AGUSTÍN RASCÓN CHU,[1] JORGE ALBERTO MÁRQUEZ ESCALANTE,[1] KARLA MARTÍNEZ-ROBINSON,[1] ALMA CAMPA-MADA,[1] and ELIZABETH CARVAJAL-MILLAN[1]

[1]*Research Center for Food and Development (CIAD, A.C), Carretera Gustavo Enrique Astizaran Rosas No. 46, Hermosillo, Sonora – 83304, Mexico, E-mail: ecarvajal@ciad.mx (E. Carvajal-Millan)*

[2]*University of Sonora, Rosales y Blvd, Luis D. Colosio, Hermosillo, Sonora – 83000, Mexico*

## ABSTRACT

Maize processing generates several co-products as if distillers dried grains with solubles (DDGS) where maize bran denotes an important fraction and represents a source of arabinoxylans (AX). The objective of this research was to extract gelling AX from DDGS and to investigate the rheological and microstructural characteristics of the gels formed. AX presented a yield of 4.2% (w/w dry basis), an arabinose to xylose ratio of 0.61, a ferulic acid (FA), diferulic acid, and triferulic acid content of 0.54, 0.10 and 0.01 µg/mg AX, respectively, and a Fourier transform infrared (FT-IR) spectrum typical of AX. The intrinsic viscosity $[\eta]$ and viscosimetric mass values for AX were 280 mL/g and 206 kDa, respectively. AX solution at 2% (w/v) formed gels induced by laccase as cross-linking agent. Cured AX gels registered storage (G') and loss (G") modulus values of 77 and 0.3 Pa, respectively, and a diferulic acid content of 0.20 µg/mg AX, only traces of triferulic acid were detected. Scanning electron microscopy (SEM) analysis of the lyophilized AX gels showed that this material resembles that of an imperfect honeycomb.

## 10.1 INTRODUCTION

Bioethanol production is increasing rapidly around the world, especially in the United States. Currently, maize is the main source for bioethanol production in this country. One of the major co-products of the maize ethanol industry is the distillers dried grains with solubles (DDGS) fraction, which consists of all the non-fermentable portions of the maize kernel, such as the bran, germ, and endosperm protein [1]. Commonly, DDGS are used as feed ingredients in animal production and sold for a very low price. In order to reduce the cost of bioethanol production, the development of value-added products from DDGS has been of interest in recent years. One of the principal components of DDGS is dietary fiber, in the form of arabinoxylans (AX). AX are non-starch polysaccharides from the cell walls of cereal endosperm [2], constituted of a linear beta-$(1{\rightarrow}4)$-xylopyranose backbone and alpha-L-arabinofuranose residues as side chains [3]. AX can present some of the arabinose residues ester-linked on (O)-5 to ferulic acid (FA) (3-methoxy, 4 hydroxycinnamic acid) [4]. Dehydrodimers of ferulic acid (di-FA) may serve to cross-link cell wall polymers and contribute to the mesh-like network of the cell wall [5, 6]. AX can gel by covalent cross-linking, involving FA oxidation by some chemical or enzymatic (laccase/$O_2$ and peroxidase/$H_2O_2$ system) free radical-generating agents [7–9]. Five di-FA structures (5-5′, 8-5′ benzo, 8-O-4′, 8-5′and 8-8′) are identified in gelled AX, the 8-5′ forms being generally preponderant [7, 10–12]. The involvement of a trimer of FA (4-O-8′, 5′-5′-dehydrotriferulic acid) in laccase cross-linked wheat or maize bran AX has been reported [10, 13]. In addition to covalent cross-links (di-FA, tri-FA), the involvement of physical interactions between AX chains was suggested to contribute to the polysaccharide gelation and gel properties [12, 13]. AX gels have an interesting technological potential as they are mostly stabilized by covalent linkages, which make them stable upon heating [3, 12]. The purpose of this research was to extract gelling AX from DDGS and to investigate the rheological and microstructural characteristics of the gels formed.

## 10.2    MATERIALS AND METHODS

### 10.2.1    MATERIALS

DDGS was kindly provided by an agro-industrial company located in Northern Mexico. Laccase (benzenediol:oxygen oxidoreductase, E.C.1.10.3.2) from *Trametes Versicolor* and other chemical products were purchased from Sigma Aldrich Co. (St. Louis, MO, USA).

### 10.2.2    AX EXTRACTION

AX were extracted from 15 Kg of DDGS according to Carvajal-Millan et al. [13] using 1 h of alkaline treatment. AX were precipitated in 60% (v/v) ethanol for 12 h at 4°C. Afterwards, ethanol was removed from AX by decantation followed by evaporation at 25°C for 12 h. Finally, AX were frozen at −20°C and lyophilized at −37°C/0.133 mbar overnight in a Freezone 6 freeze drier (Labconco, Kansas, MO, USA).

### 10.2.3    AX CHARACTERIZATION

#### 10.2.3.1    NEUTRAL SUGARS

Neutral sugars content in AX were determined according to Carvajal-Millan et al. [14], AX were subjected to hydrolysis using 2 N trifluoroacetic acid (120°C, 2 h). The reaction was stopped using an ice bath. The extracts were evaporated under air at 40°C and then rinsed twice with 200 µL of water. The evaporated extracts were solubilized in 500 µL of water. Mannitol was used as an internal standard. Samples were filtered (0.45 µm filter, Whatman) and analyzed by high performance liquid chromatography (HPLC) using a Supelcogel Pb column (300 × 7.8 mm; Supelco, Inc., Bellefonte, PA, USA), eluted with 5 mM $H_2SO_4$ (filtered 0.2 µm, Whatman) at 0.6 mL/min and 50°C. A refractive index detector Varian 2414 (Varian, St. Helens, Australia) was used.

#### 10.2.3.2    PROTEIN

Protein content in AX was determined according to the Dumas method [15], using a Leco-FP 528 nitrogen analyzer.

### 10.2.3.3  PHENOLIC ACIDS

FA, di-FA, and tri-FA contents in AX powder and gels were analyzed by RP-HPLC after their de-esterification as described by Vansteenkiste et al. [12] and Rouau et al. [16]. An Altima C18 column (250 x 4.6 mm; Alltech Associates, Inc., Deerfield, IL, USA) and a photodiode array detector Waters 996 (Millipore Co., Milford, MA, USA) were used. Detection was followed by UV absorbance at 320 nm. A gradient elution was done using acetonitrile and sodium acetate buffer (0.05 M, pH 4.0) at 1 mL/min at 35°C, in linear gradients from 15/85 to 35/65 in 30 min; from 35/65 to 60/40 in 0.5 min; from 60/40 to 15/85 in 4.5 min, and finally maintained at 15/85 for 5 min.

### 10.2.3.4  INTRINSIC VISCOSITY AND VISCOSIMETRIC MOLECULAR WEIGHT (MW)

Viscosity measurements were made by determination of the flow times of AX solutions in water (from 0.06 to 0.1% w/v). An Ubbelohde capillary viscometer at $25 \pm 0.1$ °C, immersed in a temperature-controlled water bath was used. The intrinsic viscosity [η] was estimated from relative viscosity measurements (ηrel) of AX solutions by extrapolation of Kraemer and Mead and Fouss curves to "zero" concentration [14]. The viscosimetric molecular weight (Mv) was calculated from the Mark-Houwink relationship, $Mv = ([\eta]/k)^{1/\alpha}$.

### 10.2.3.5  FOURIER TRANSFORM INFRARED (FT-IR) SPECTROSCOPY

FT-IR spectrum of dry AX was recorded on a Nicolet FT-IR spectrophotometer (Nicolet Instrument Corp. Madison, WI, USA). The samples were pressed into KBr pellets (2 mg/200 mg KBr). A blank KBr disk was used as background. The spectra were measured in absorbance mode from 400–4000 $cm^{-1}$.

### 10.2.4  AX GELS

AX solution (2% w/v) was prepared in 0.05 M citrate-phosphate buffer, pH 5.5. Laccase (1.675 nkat per mg AX) was added to solutions as a cross-linking agent for the formation of gels. Gels were allowed to develop for 1 h at 25°C.

## 10.2.4.1   RHEOLOGY

Small amplitude oscillatory shear was used to follow the gelation process of AX solution. AX solution was mixed with laccase (1.675 nkat per mg AX) and immediately placed on the parallel-plate geometry (40 mm in diameter) of a strain-controlled rheometer (Discovery HR-2 rheometer; TA Instruments, New Castle, DE, USA). Exposed edges were recovered with silicone to prevent evaporation. The dynamic rheological parameters used to evaluate the gel network were the storage modulus (G′), loss modulus (G") and crossover point (G′>G"). AX gelation was monitored at a frequency of 0.25 Hz and 5% strain. At the end of the network formation, a frequency sweep (0.01–10 Hz) was carried out. The rheological measurements were performed in duplicate [14].

## 10.2.4.2   MICROSTRUCTURE

AX gels were frozen at −20°C and lyophilized at −37°C/0.133 mbar overnight in a Freezone 6 freeze drier (Labconco, Kansas, MO, USA). The superficial and internal morphology of freeze-dried AX gel was studied by scanning electron microscopy (SEM) (JEOL 5410LV, JEOL, Peabody, MA, USA) at low voltage (20 kV). SEM image was obtained in secondary electrons image mode.

## 10.2.5   STATISTICAL ANALYSIS

Chemical determinations were done in duplicates and the coefficients of variation were lower than 5%. Rheological measurements were done in duplicate, and the coefficients of variation were lower than 8%. All results are expressed as mean values.

## 10.3   RESULTS AND DISCUSSION

### 10.3.1   EXTRACTION AND CHARACTERIZATION OF AX

AX were extracted from 15 Kg of DDGS, giving a yield of 4.2% (w AX/w DDGS) on a dry matter basis, which is in the range reported by Carvajal-Millan et al. [13] using a maize bran co-product recovered from maize flour industry. AX from cereal bran are water unextractable (WUAX) as they present high Mw and FA content, however, after alkaline treatment under controlled conditions as those used in the present study, WUAX from maize bran can be extracted in water. It is important to mention that AX present in cereal endosperm are water-extractable (WEAX); however, the extraction of WEAX is more expensive than that of the AX used in the present study because the use of enzymes is required, extraction yields are lower (0.5% w WEAX/w cereal endosperm) and the cereal endosperm is usually destined to human consumption, as for example wheat flour [7, 17]. On the contrary, in the present investigation, the extraction did not involve enzymes, the yield is higher (4.2% w AX/w DDGS) and the source is maize DDGS from bioethanol plants.

AX composition is presented in Table 10.1. The arabinose + xylose content in the AX extracted from DDGS was 69% db, which is close to the value reported for AX recovered from other maize co-product [18]. The arabinose to xylose ratio (A/X) was 0.61, indicating a moderately branched structure similar to that earlier reported in maize bran AX [19]. Small levels of glucose, galactose, and mannose were also quantified. The FA content (0.54 µg/mg AX) was higher than that reported for AX isolated from a maize co-product generated during nixtamalization (cooking maize grains in a lime solution) (0.012–0.23 µg/mg AX) [18, 20]. The higher FA content in AX from DDGS in relation to AX isolated from nixtamalized maize co-products can be due to the partial de-esterification of FA attached to AX chains occurring during maize grains cooking under basic conditions while during maize bioethanol production no alkaline treatment is involved allowing a better preservation of FA in the AX molecules. The di-FA and tri-FA content in AX was 0.10 and 0.01 µg/mg AX, respectively. The presence of di-FA and tri-FA in AX from different cereals has been previously reported and related to the possible presence of cross-linked AX chains [10, 13, 21]. The intrinsic viscosity [η] and viscosimetric mass values for AX were 280 mL/g and 206 kDa, respectively, which are in the range reported for maize AX [13].

The molecular identity of AX was analyzed by FT-IR spectroscopy (Figure 10.1). The FT-IR spectrum showed a broad absorbance region for polysaccharides at 1200–800 cm$^{-1}$ [22, 23]. AX presented a typical spectrum of AX with the maximum absorption band at 1158 cm$^{-1}$, which could be related to the antisymmetric C-O-C stretching mode of the β-(1-4) glycosidic linkages between the sugar units [24]. The bands at 3413 cm$^{-1}$ was associated to the OH stretching and the band at 2854 cm$^{-1}$ was related to the CH$_2$ groups [25].

**TABLE 10.1**    Composition of AX (g/100 g)

| | |
|---|---|
| Arabinose[a] | 28.81 ± 1.30 |
| Xylose[a] | 41.70 ± 1.95 |
| Glucose[a] | 3.36 ± 0.31 |
| Galactose[a] | 4.20 + 0.06 |
| Mannose[a] | 0.96 + 0.07 |
| Protein[a] | 3.20 + 0.05 |
| Ferulic acid[b] | 0.54 + 0.06 |
| Diferulic acids[b] | 0.10 + 0.02 |
| Triferulic acid[b] | 0.01 ± 0.01 |

[a] *Results are expressed in g/100 g AX dry matter.*
[b] *Phenolics are expressed in µg/mg AX dry matter.*
*All results are obtained from duplicates.*

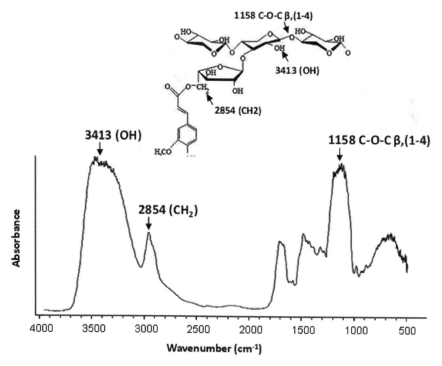

**FIGURE 10.1**  Fourier transform infrared spectra (FTIR). The arrows indicate the characteristic absorption bands of AX.

## 10.3.2   AX GELATION AND GELS CHARACTERISTICS

The cross-linking process of the AX was rheologically investigated by small-amplitude oscillatory shear. The kinetics of gelation and the mechanical spectrum of the AX were registered. Storage (G') and loss (G") modulus as a function of time are presented in Figure 10.2. The gelation profile exhibited an initial increase in G' followed by a stability region (*plateau*). The rapid increase in G' occurring before the *plateau* region is attributed to the initial formation of covalent linkages between FA residues of adjacent AX chains. The higher G' value observed for AX gel was 77 Pa. The rheological measurements showed that gelation times ($t_g$), calculated from the crossover of the G' and G" curves (G' > G") was 8 min. The $t_g$ value indicates the sol/gel transition point and at this point G' is equal to G" [26]. The mechanical spectra of AX gels (Figure 10.3) after 1 h of gelation exhibited a typical solid-like material behavior with a linear G' independent of frequency and G" much smaller than G' and dependent on frequency [27]. Similar behavior has been previously reported for other maize AX gels [28–30]. The di-FA structures present in AX before gelation were 5′-5′ and 8-5′ at 41 and 59%, respectively (Figure 10.4(A)). The 8-O-4' and 8-8' dehydrodimer were not detected. The presence of 5′-5′ and 8-5' di-FA structures in maize AX has been indicated in earlier investigations [13, 20]. After AX gelation, the 5-5' di-FA content decreased down to 11% and that of 8-5'dimer increased up to 89% (Figure 10.4(B)). The predominance of 8-5' dimers has been also observed in wheat, barley, rye, and triticale AX [11, 31]. The decrease in 5′-5′ di-FA content after AX gelation could favor the inter-polysaccharide chain bonds [32] allowing the development of an effective polymeric network structure. AX cross-linking promotes the selective fermentation of this polysaccharide in the colonic region, limiting the growth of bacteria that are not considered beneficial (Bacteroides) and favoring the growth of probiotics such as bifidobacteria [33]. In this regard, the formation of an AX gel with effective interchain di-FA structures would improve the functionality of these gels and their potential use in the food industry [34].

SEM was used to investigate the surface and internal morphology of AX gel. In Figure 10.5, it can be seen that the AX gel surface present a slightly rough surface while the gel interior reveals a heterogeneous porous microstructure that resembles an imperfect honeycomb type, which agrees with previous reports on maize AX gels [18, 22, 35]. The microstructural characteristics of polysaccharide gels are determined by the molecule characteristics and concentration, as well as the conditions of preparation and drying of the gel. In this last aspect, the process of ice formation and sublimation that lyophilization implies is determinant in the morphology of the material.

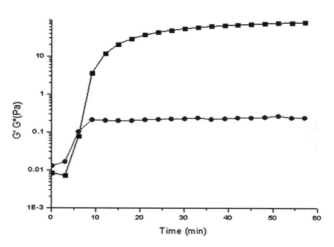

**FIGURE 10.2** Rheological kinetics of 2% (w/v) AX solution during gelation by laccase at 0.25 Hz, 5% strain and 25°C.

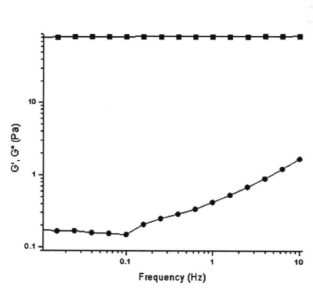

**FIGURE 10.3** Mechanical spectrum of 2% (w/v) AX gel. Rheological measurements at 25°C and 5% strain.

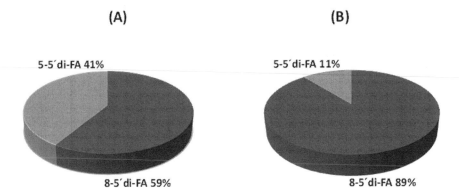

**FIGURE 10.4**    di-FA structures in AX before (A) and after (B) gel formation induced by laccase.

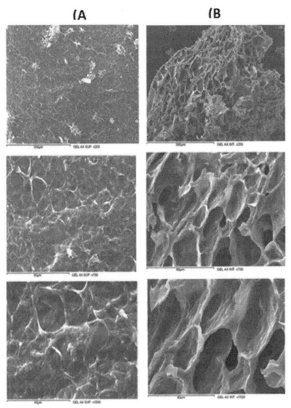

**FIGURE 10.5**    Scanning electron microscopy images of lyophilized AX gel surface (A) and interior (B) at 200× (A),700× (B), and 1500× (C).

## 10.4   CONCLUSION

Gelling AX can be recovered from maize DDGS having a yield and FA content higher than those have reported by using other maize co-products. The gels formed at 2% (w/v) in AX presented a storage modulus of 77 Pa and an imperfect honeycomb type microstructure, which are representatives of this kind of gels. Extraction of AX from DDGS may represent an opportunity in sustainable bioethanol waste management and utilization through upcycling to value-added products.

## ACKNOWLEDGMENTS

This work was supported by the "Fund to support research on the Sonora-Arizona region 2019," Mexico [Grant to E. Carvajal-Millan].

## KEYWORDS

- arabinoxylans
- dehydrodimers of ferulic acid
- distillers dried grains with solubles
- ferulic acid
- gelation
- maize
- microstructure

## REFERENCES

1. Du, C., Campbell, G. M., Misailidis, N., Materos-Salvador, F., Sadhukhan, J., Mustafa, M., & Weightman, R. M., (2009). Evaluating the feasibility of commercial arabinoxylan production in the context of a wheat biorefinery principally producing ethanol. Part 1: Experimental studies of arabinoxylan extraction from wheat bran. *Chemical Engineering Research and Design, 87*, 1232–1238. https://doi.org/10.1016/j.cherd.2008.12.027.
2. Fincher, G. B., Sawyer, W. H., & Stone, B. A., (1974). Chemical and physical properties of an arabinogalactan-peptide from wheat endosperm. *Biochemical Journal, 139*(3), 535–545.

3. Izydorczyk, M. S., & Biliaderis, C. G., (1995). Cereal arabinoxylans: advances in structure and physicochemical properties. *Carbohydrate Polymers, 28*, 33–48. https://doi.org/10.1016/0144-8617(95)00077-1.

4. Smith, M. M., & Hartley, R. D., (1983). Occurrence and nature of ferulic acid substitution of cell-wall polysaccharides in graminaceous plants. *Carbohydrate Research 118*, 65–80. https://doi.org/10.1016/0008-6215(83)88036-7.

5. Bunzel, M., Ralph, J., Brüning, P., & Steinhart, H., (2006). Structural identification of dehydrotriferulic and dehydrotetraferulic acids isolated from insoluble maize bran fiber. *Journal of Agricultural and Food Chemistry, 54*(17), 6409–6418. https://doi.org/10.1021/jf061196a.

6. Iiyama, K., Lam, T. B. T., & Stone, B. A., (1994). Covalent cross-links in the cell wall. *Plant Physiology, 104*(2), 315.

7. Figueroa-Espinoza, M. C., & Rouau, X., (1998). Oxidative cross-linking of pentosans by a fungal laccase and horseradish peroxidase: Mechanism of linkage between feruloylated arabinoxylans. *Cereal Chemistry, 75*(2), 259–265. https://doi.org/10.1094/CCHEM.1998.75.2.259.

8. Geissmann, T., & Neukom, H., (1973). On the composition of the water-soluble wheat flour pentosans and their oxidative gelation. *Lebensmittel-Wissenschaft Und -Technologie, 6*, 59–72.

9. Hoseney, R. C., & Faubion, J. M., (1981). A mechanism for the oxidative gelation of wheat flour water-soluble pentosans. *Cereal Chemistry*.

10. Carvajal-Millan, E., Guigliarelli, B., Belle, V., Rouau, X., & Micard, V., (2005a). Storage stability of laccase induced arabinoxylan gels. *Carbohydrate Polymers, 59*, 181–188. https://doi.org/10.1016/j.carbpol.2004.09.008.

11. Schooneveld-Bergmans, M. E. F., Dignum, M. J. W., Grabber, J. H., Beldman, G., & Voragen, A. G. J., (1999). Studies on the oxidative cross-linking of feruloylated arabinoxylans from wheat flour and wheat bran. *Carbohydrate Polymers, 38*(4), 309–317. https://doi.org/10.1016/S0144-8617(98)00121-0.

12. Vansteenkiste, E., Babot, C., Rouau, X., & Micard, V., (2004). Oxidative gelation of feruloylated arabinoxylan as affected by protein. Influence on protein enzymatic hydrolysis. *Food Hydrocolloids, 18*, 557–564. https://doi.org/10.1016/j.foodhyd.2003.09.004.

13. Carvajal-Millan, E., Rascón-Chu, A., Márquez-Escalante, J. A., Micard, V., De León, N. P., & Gardea, A., (2007). Maize bran gum: Extraction, characterization and functional properties. *Carbohydrate Polymers, 69*, 280–285. https://doi.org/10.1016/j.carbpol.2006.10.006.

14. Carvajal-Millan, E., Landillon, V., Morel, M. H., Rouau, X., Doublier, J. L., & Micard, V., (2005b). Arabinoxylan gels: Impact of the feruloylation degree on their structure and properties. *Biomacromolecules, 6*, 309–317. http://dx.doi.org/10.1021/bm049629a.

15. AOAC, (1990). *In Official Methods of Analysis, of the Association of Official Analytical Chemists*. The Association, Virginie, USA.

16. Rouau, X., Cheynier, V., Surget, A., Gloux, D., Barron, C., Meudec, E., Montero, J. L., & Criton, M. A., (2003). A dehydrotrimer of ferulic acid from maize bran. *Phytochemistry, 63*, 899–903. https://doi.org/10.1016/S0031-9422(03)00297-8.

17. Dervilly, G., Saulnier, L., Roger, P., & Thibault, J. F., (2000). Isolation of homogeneous fractions from wheat water-soluble arabinoxylan. Influence of the structure on their

macromolecular characteristics. *Journal of Agricultural and Food Chemistry, 48*, 270–278. doi: 10.1021/jf990222k.

18. Paz-Samaniego, R., Carvajal-Millan, E., Sotelo-Cruz, N., Brown, F., Rascón-Chu, A., López-Franco, Y. L., & Lizardi-Mendoza, J., (2016). Maize processing wastewater upcycling in Mexico: Recovery of arabinoxylans for probiotic encapsulation. *Sustainability, 8*(11), 1104. https://doi.org/10.3390/su8111104.

19. Singh, V., Doner, L. W., Johnston, D. B., Hicks, K. B., & Eckhoff, S. R., (2000). Comparison of coarse and fine corn fiber for corn fiber gum yields and sugar profiles. *Cereal Chemistry, 77*, 560–561. http://dx.doi.org/10.1094/CCHEM.2000.77.5.560.

20. Niño-Medina, G., Carvajal-Millán, E., Rascón-Chu, A., Lizardi, J., Márquez-Escalante, J., Gardea, A., Martínez-López, A. L., & Guerrero, V., (2009). Maize processing wastewater arabinoxylans: Gelling capability and cross-linking content. *Food Chemistry, 115*(4), 1286–1290. doi: 10.1016/j.foodchem.2009.01.046.

21. Bunzel, M., Ralph, J., Marita, J. M., Hatfield, R. D., & Steinhart, H., (2001). Diferulates as structural components in soluble and insoluble cereal dietary fiber. *Journal of the Science of Food and Agriculture, 81*, 653–660. https://doi.org/10.1002/jsfa.861.

22. Iravani, S., Fitchett, C. S., & Georget, D. M. R., (2011). Physical characterization of arabinoxylan powder and its hydrogel containing a methylxanthine. *Carbohydrate Polymers, 85*, 201–207. https://doi.org/10.1016/j.carbpol.2011.02.017.

23. Li, L., Ma, S., Fan, L., Zhang, C., Pu, X., Zheng, X., & Wang, X., (2016). The influence of ultrasonic modification on arabinoxylans properties obtained from wheat bran. *International Journal of Food Science and Technology, 51*(11), 2338–2344. doi: 10.1111/ijfs.13239.

24. Barron, C., & Rouau, X., (2008). FTIR and Raman signatures of wheat grain peripheral tissues. *Cereal Chemistry, 85*, 619–625. https://doi.org/10.1094/CCHEM-85-5-0619.

25. Urias-Orona, V., Huerta-Oros, J., Carvajal-Millán, E., Lizardi-Mendoza, J., Rascón-Chu, A., & Gardea, A. A., (2010). Component analysis and free radicals scavenging activity of *Cicer arietinum* L. husk pectin. *Molecules, 15*, 6948–6955. https://doi.org/10.3390/molecules15106948.

26. Doublier, J. L., & Cuvelier, G., (1996). Gums and hydrocolloids: Functional aspects. In: Eliasson, A. C., (ed.), *Carbohydrates in Food* (pp. 283–318). New York: Marcel Dekker.

27. Ross-Murphy, S. B., & Shatwell, K. P., (1993). Polysaccharide strong and weak gels. *Biorheology, 30*, 217–227. doi: 10.3233/BIR-1993-303-407.

28. Kale, M. S., Hamaker, B. R., & Campanella, O. H., (2013). Alkaline extraction conditions determine the gelling properties of corn bran arabinoxylans. *Food Hydrocolloids, 31*, 121–126. https://doi.org/10.1016/j.foodhyd.2012.09.011.

29. Berlanga-Reyes, C. M., Carvajal-Millan, E., Lizardi-Mendoza, J., Rascón-Chu, A., Marquez-Escalante, J. A., & Martinez-Lopez, A. L., (2009). Maize arabinoxylan gels as protein delivery matrices. *Molecules, 14*, 1475–1482. https://doi.org/10.3390/molecules14041475.

30. Carvajal-Millan, E., Guilbert, S., Doublier, J. L., & Micard, V., (2006). Arabinoxylan/protein gels: Structural, rheological and controlled release properties, *Food Hydrocolloids, 20*, 53–61. https://doi.org/10.1016/j.foodhyd.2005.02.011.

31. Dervilly-Pinel, G., Rimsten, L., Saulnier, L., Andersson, R., & Åman, P., (2001). Water-extractable arabinoxylan from pearled flours of wheat, barley, rye and triticale. Evidence for the presence of ferulic acid dimers and their involvement in gel formation. *Journal of Cereal Science, 34*(2), 207–214. https://doi.org/10.1006/jcrs.2001.0392.

32. Hatfield, R. D., & Ralph, J., (1999). Modeling the feasibility of intramolecular dehydrodiferulate formation in grass walls. *J. Sci. Food Agric., 79*(3), 425–427. https://doi.org/10.1002/(SICI)1097-0010(19990301)79:3<425::AID-JSFA282>3.0.CO;2-U.

33. Hopkins, M. J., Englyst, H. N., Macfarlane, S., Furrie, E., Macfarlane, G. T., & McBain, A. J., (2003). Degradation of cross-linked and non-cross-linked arabinoxylans by the intestinal microbiota in children. *Appl. Environ. Microbiol., 69*, 6354–6360. https://doi.org/10.1128/AEM.69.11.6354-6360.2003.

34. Martínez-López, A. L., Carvajal-Millan, E., Marquez-Escalante, J., Campa-Mada, A. C., Rascón-Chu, A., López-Franco, Y. L., & Lizardi-Mendoza, J., (2018). Enzymatic cross-linking of ferulated arabinoxylan: Effect of laccase or peroxidase catalysis on the gel characteristics. *Food Science and Biotechnology*. https://doi.org/10.1007/s10068-018-0488-9.

35. Martínez-López, A. L., Carvajal-Millan, E., Rascón-Chu, A., Márquez-Escalante, J., Martínez-Robinson, K., (2013). Gels of ferulated arabinoxylans extracted from nixtamalized and non-nixtamalized maize bran: Rheological and structural characteristics. *CyTA-Journal of Food 11*(S1), 22–28. https://doi.org/10.1080/19476337.2013.781679.

# CHAPTER 11

# Cereal Arabinoxylans Gelled Particles: Advances in Design and Applications

YUBIA B. DE ANDA-FLORES,[1] ELIZABETH CARVAJAL-MILLAN,[1] AGUSTÍN RASCÓN-CHU,[2] ALMA CAMPA-MADA,[1] JAIME LIZARDI-MENDOZA,[1] JUDITH TANORI-CORDOVA,[3] and ANA L. MARTÍNEZ-LÓPEZ[4]

[1]Biopolymers-CTAOA, Research Center for Food and Development (CIAD, A.C.), Carretera Gustavo Enrique Astiazarán Rosas No. 46, Hermosillo – 83304, Sonora, Mexico, Tel.:+52-662-289-2400, E-mail: ecarvajal@ciad.mx (E. Carvajal-Millan)

[2]Biotechnology-CTAOV. Research Center for Food and Development (CIAD, A.C.), Carretera Gustavo Enrique Astiazarán Rosas No. 46, Hermosillo – 83304, Sonora, Mexico

[3]Department of Polymers and Materials Research, University of Sonora, Hermosillo – 83000, Sonora, Mexico

[4]NANO-VAC Research Group, Department of Chemistry and Pharmaceutical Technology, University of Navarra, Pamplona – 31008, Spain

## ABSTRACT

Arabinoxylans (AX) are non-starch polysaccharides, which are part of the hemicellulose fraction in the cell wall of cereal tissues such as pericarp and endosperm. The AX physicochemical properties, such as molecular weight, xylan backbone substitution (A/X), and ferulic acid content (FA), contribute to its ability to form covalent gels. AX gelled particles can act as delivery systems. Besides, these particles can be considered health-promoting agents for biomedical, pharmaceutical, and nutritional applications. This chapter provides a review of AX gelled particles (macro, micro, and nanoparticles) characteristics and potential applications not only as simple carriers but as bioactive delivery systems.

## 11.1  INTRODUCTION

Particles based on polysaccharides have come into interest as a novel alternative for delivery systems [1]. Macro, micro, and nanoparticle delivery systems can improve bioactive agent stability [2]. Examples of polysaccharides studied for this purpose are chitosan, alginate, and dextran. Because of their biodegradability, biocompatibility, nontoxicity, and hydrophilic, polysaccharide particles are attractive delivery systems [3, 4]. Polysaccharide gels have been studied as possible delivery systems for bioactive ingredients due to their enhanced biocompatibility and capacity to retain water [4, 5]. Most polysaccharide gels are affected by the acid pH of the stomach, which promotes the early release of the drug [6, 7]. Among polysaccharides, ferulated arabinoxylans can form covalent gels that can resist the conditions of the human upper gastrointestinal system [8].

The non-invasive delivery systems for the oral administration of bioactive agents, polymeric particles, and liposomes have been investigated [1, 9]. It should be noted that protein structures such as enzymes cannot be administered by oral route because they would break down in the digestive process, by the acidic pH of the stomach, and by the proteases of the stomach and small intestine [10]. The colonic region has lower proteolytic activity compared to the small intestine; therefore, the small intestine has been considered as a possible absorption site for peptides and proteins administered orally [6, 11].

AX are non-starch polysaccharides commonly found in the cell wall, outer layer, and endosperm of cereals. AX polymeric chain consists of xylose in $\beta$-1,4 with ramifications of $\alpha$-L-arabinofuranose in $\alpha$-1,3 and $\alpha$-1,2. Arabinose can be esterified with monomeric or dimeric ferulic acid (FA); thus, AX can be crosslinked [5, 12, 13]. AX can form covalent gels by oxidative coupling of ferulic residues resulting in the formation of dimers (di-FA) (5-5', 8-5' benzo, 8-O-4,' 8-5' and 8-8' isomers) and a trimer (tri-FA) (4-O-8,' 5'-5''-dehydrotriferulic acid). The physical interactions between the AX chains contribute to the crosslinking process defining the characteristics of the gel [14].

Polysaccharide particles have been promoted as delivery systems because of their biocompatibility [2]. These particles improve delivery and specificity

[16]. The purpose of encapsulation as a delivery vehicle is to maintain the stability and functionality of the active ingredients and protect them from external factors or extreme conditions that may affect those [17]. Therefore, oral administration of bioactive-loaded particles can provide stability against enzymatic degradation in the gastrointestinal environment, due to its protective polymer matrix [2]. Generally, particles can be made from almost any chemically stable structure, most of them derived from materials such as synthetic and natural polymers [18, 19]. Different particle sizes can be based on polysaccharides such as alginate, agar, agarose, and chitosan, among others [2, 16]. Nanoparticles are generally found on a nanometer scale from 1–100 nm. However, biopolymeric nanoparticles present size between 10–1000 nm [20]. In general, polysaccharide particles offer advantages such as biocompatibility and biodegradability [21]. The development of polysaccharide particles as delivery systems has been of interest because these materials can effectively carry molecules or cells to target sites decreasing side effects [22–29]. Hydrophilic groups in polysaccharides favor the molecule interaction with biological tissues, increasing biological properties such as mucoadhesion [30].

## 11.2   ARABINOXYLANS (AX)

The AX is mainly found in cereal grains such as wheat, rice, maize, rye, oats, barley, millet, and sorghum. In general, they are non-starch polysaccharides that are part of a group of hemicelluloses present in the cell wall of cereal tissues, such as the pericarp and endosperm [12, 14, 31, 32]. Due to this, it is essential to know AX source, chemical structure, and physicochemical characteristics, degree of ferulation and branching, molecular weight (Mw), and viscosity.

### 11.2.1   SOURCES

It is known that AX are obtained mainly from different types of cereals such as wheat, maize, rice, barley, oats, rye, and sorghum, which have been widely studied [12]. However, the content of AX varies depending on genotypic and environmental differences [32]. In the cell wall of cereal grains, most AX and other polymers are crosslinked with other cell wall components (cellulose microfibrils) by hydrogen bonds, which confers specific stability [12, 13]. AX can be classified as water-extractable (WEAX) and water-unextractable

(WUAX) fractions. WEAX are part of the endosperm; WUAX are extracted from the pericarp, through chemical or enzymatic treatments [12, 33]. Commonly, the extraction processes involve the inactivation of endogenous enzymes in an aqueous environment; besides, hydrolytic enzymes are used to eliminate proteins and starch extract, which favors obtaining AX with higher purity [14].

In the literature, it is mentioned that maize (*Zea mays* L.) represents an essential source of food for humans and many animals [34]. Maize grains represent 15–56% of the dietary caloric intake in many developed and under-developed countries. In Mexico, the production of maize grain during the 2016–2017 autumn-winter agricultural cycle was 7 million 866 tons [35].

In the nixtamalization process of maize, the nejayote is obtained as a by-product, from which the AX are extracted [36]. Niño-Medina et al. [37] reported the lowest performance with a percentage of 8% (w/w) of AX. Paz-Samaniego et al. [38] also reported a yield of 0.90% (w/w) of nejayote AX. Maize pericarp derived from the industrial dry milling process in Mexico is another source of AX. It is considered an effective and economical source with a yield of 25–30% [39, 40]. Martínez-López [41] reported an extraction yield of maize pericarp AX of 18% while Morales-Burgos et al. [42] obtained a lower yield of 1.4% (w/w) for the same source. Another maize by-product derived from the distillation process is the distillers dried grains with solubles (DDGS) and can be used as a source of AX [43]. Méndez-Encinas [44] reported a DDGS AX yield of 2.5% (w/w), a lower than the one obtained for nixtamalization and dry grinding.

## 11.2.2  CHEMICAL STRUCTURE AND PHYSICOCHEMICAL CHARACTERISTICS

AX is composed of a linear chain of xylose in $\beta$-(1→4) with branches of $\alpha$-L-arabinofuranose in $\alpha$-(1→3) and $\alpha$-(1→2). The arabinose can be esterified at the (O)-5 position to monomeric or dimeric FA; hence the AX can be crosslinked (Figure 11.1) [14, 45]. The chains of FA can be linked to $\beta$-glucans, cellulose, glucose, and protein. In cereal bran, the difference occurs in the arabinose residue substitution in the xylan chain, in the relative proportions and sequences of several linkages between xylose and arabinose, and the presence of other substituents [12].

The physicochemical properties depend directly on the chemical structure of the AX. The AX contains in their chemical structure functional groups, such as hydroxyl, carboxyl, benzene, and esters. They provide specific physicochemical characteristics that are related to Mw, viscosity, and solubility [45].

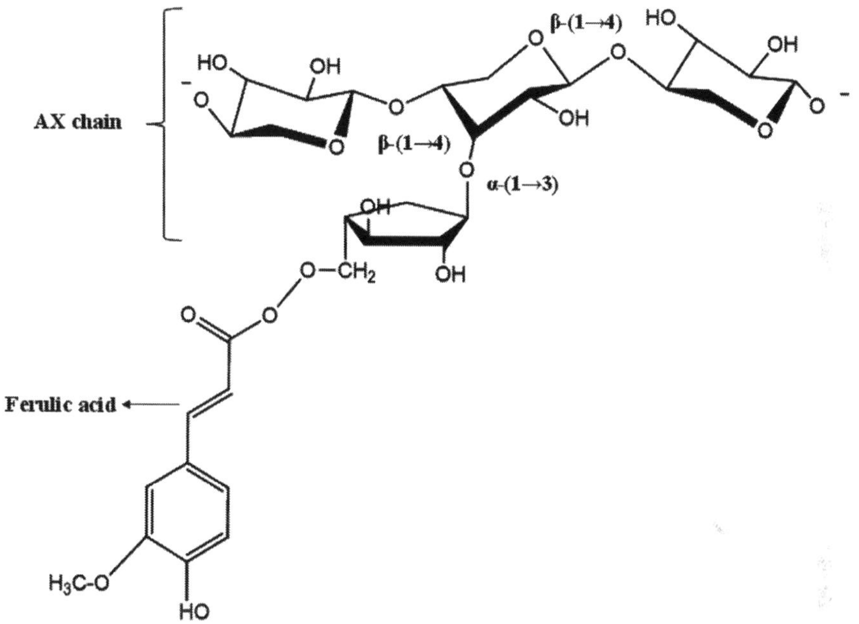

**FIGURE 11.1**   Chemical structure of a fraction of ferulated arabinoxylans [14, 45].

## 11.2.3   DEGREE OF FERULATION

The polysaccharide chains of AX can be crosslinked through dimerization reactions under oxidative conditions forming dimers [di-FA (5-5,' 8-5' benzo, 8-O-4,' 8-5' and 8-8')]; the most abundant di-FA is the 8-5' form [46]. The formation of a tri-FA (4-O-8,' 5'-5"-dehydrotriferulic acid) has also been reported [14, 39]. In addition to covalent crosslinks (di-FA, tri-FA), the involvement of physical interactions between AX chains was suggested to contribute to gelation and gel properties of the polysaccharide [39, 47].

Carvajal-Millan et al. [39] reported in maize bran gum AX a FA content of 0.34 μg/mg, and di-FA and tri-FA in concentrations of 0.77 and 0.39 μg/mg, respectively. These data are indicative of some chains of AX that may be interred or intra-cross linked. Niño-Medina et al. [48], Martínez-López [41] and Paz-Samaniego et al. [38] reported low FA contents in AX extracted from residues of maize nixtamalization process (0.23 µg/mg, 0.25 µg/mg and 0.012 µg/mg, respectively). AX extracted from bioethanol production from maize (DDGS) presented a FA content of 6.05 µg/mg [44], and AX extracted from the maize dry milling process have an AF content of 7.18 µg/mg [42]. In a more recent study, Astiazarán-Rascón [15] reported a FA content of 12 µg/mg in maize pericarp AX.

### 11.2.4   DEGREE OF BRANCHING

The degree of branching, degree of substitution, or proportion of arabinose/xylose (A/X) refers to the arabinoses along the main chain of xyloses in the AX. The arabinose-to-xylose (A/X) ratio can vary from 0.3 to 1.1, depending on the cereal and the cereal tissue where the AX is located [14, 49]. The A/X ratio indicates whether a moderately structure or highly branched [39]. Izydorczyk and Biliaderis [31] reported the relationship between arabinose and xylose (particularly A/X ratio) for AX from different cereals, from both the pericarp and endosperm. Endosperm AX has an A/X ratio of 0.48–0.83, and pericarp AX presents values of 0.57–1.07; thus, the highest proportions of A/X are found in the pericarp. Several authors report the A/X ratio of maize pericarp AX from different maize by-products. In maize pericarp AX, A/X values between 0.65 and 1.08 for moderately to highly branched structures have been reported [38, 39, 42, 44, 48].

### 11.2.5   MOLECULAR WEIGHT (MW) AND VISCOSITY

The Mw of AX varies from 10 to 10,000 kDa for WEAX and higher than 10,000 kDa for WUAX. Mw will also depend on the AX source and the method used for its determination [14, 31]. In cereals such as wheat, barley, and rye, the MW of WEAX is within 200 to 300 kDa, with a high polydispersity index of 1.7–2. The high polydispersity reflects a range of polymers exhibiting different mass and structure. In maize bran AX, MW can be affected by the extraction conditions such as time, pH, and temperature [50]. Niño-Medina et al. [48] reported a Mw of 60 kDa, while Martínez-López et al. [51] reported a Mw of 197 kDa in maize bran AX. Because of the high MW and high FA content, AX quickly form covalent and non-covalent bonds between the AX chains and with other cell wall components, such as proteins, β-glucans, lignin, and cellulose; thus, a large proportion of AX cannot be extracted in water [12].

The intrinsic viscosity ($[\eta]$) of the AX depends on MW and the structure of the polymer, and it is an indicator of the hydrodynamic volume occupied by the mass of AX. The A/X ratio has an effect on the $[\eta]$ of the AX, due to arabinose residues favoring the interaction of the polysaccharide with water [50, 52]. The $[\eta]$ of nejayote AX have an average $[\eta]$ of 183 mL/g [48]. Carvajal-Millan et al. [39] reported an $[\eta]$ of 208 mL/g of maize bran AX.

## 11.2.6 SOLUBILITY

The solubility of AX and polysaccharides, in general, is mainly determined by the molecular structure and Mw. The branching of polysaccharide chains may reduce the intermolecular association, increasing the solubility. The side chain length, the number of substituents, and their distribution may modify the solubility of the polymers. The water solubility of AX is not only related to the structural characteristics of the polymer chain but also the covalent linkage to other polymers in the wall [50]. Consequently, it can be deduced that at higher crosslinking, there is less solubility of the AX, as they present a lower exposure of their polar functional groups (e.g., hydroxyls and carboxylates).

On the contrary, at lower crosslinking content, higher exposure, and, therefore, higher solubility [12]. The A/X ratio influences the ability of the chains to aggregate, which can affect their solubility. Therefore, AX with fewer arabinose substitutions has a lower solubility in water [12, 53–55].

## 11.2.7 FUNCTIONAL PROPERTIES OF AX

The study of AX and their properties has resumed worldwide interest in the last 10 years because AX has been recognized as safe ingredient (GRAS) [56]. Its application in the food industry, pharmaceuticals, and cosmetics, among others, has aroused great interest [57, 58]. In addition, AX has the capacity to provide health benefits [59].

## 11.2.8 BENEFICIAL EFFECTS ON HEALTH

The AX has the ability to act as prebiotics, antioxidants, and antimicrobial [57, 60]. They have also been related to the reduction of serum cholesterol, regulation of blood sugar, stimulation of the immune system, and reduction of the risk of coronary diseases, among others [12]. Its properties as soluble fiber have made the AX of great interest because they help with better absorption of calcium and magnesium and reduce the risk of colon cancer [49]. AX has been considered as a human dietary fiber because the human colonic microbes ferment them when they cross the digestive system [61]. In addition, AX are expected to be sensitive to hydrolysis because the L-arabinose released is susceptible to increase up to 10% at 37°C and gastric pH 1–3 than other hemicelluloses [62]. In the human gut, the microbiota or microbiome comprise trillions of microbes with a role in human health. Their function is to regulate the physiological process as energy and fat metabolism, host immune system development, and barrier function (preventing pathogen invasion and colonization) [63]. The gut microbiota can integrate into their ecosystem oral strains of bacteria called probiotics, promoting specific changes into the gut [64]. Probiotics have been defined as 'live microorganisms that confer a health benefit on the host' and show an impact on the intestinal microbiota or improve the immune system [65]. Examples of essential intestinal strains are *Lactobacillus, Bifidobacterium* species [66].

Colon microbiota stimulates AX fermentation in the large intestine by *Bacteroides* and *Bifidobacterium.* At the same time, AX, resist digestion in the human small intestine [8]. Wheat WUAX and WEAX have been considered prebiotics because it is fermentative properties by gut microbiota [67]. Prebiotics 'certain nondigestible carbohydrates, following fermentation by bacteria, can drive qualities and selective changes in the composition of the gut microbiota, which have a beneficial effect on the host's health' [64]. The AX degradation is the result of different bacterial enzymes as endo-xylanases, beta-xylosidases, and alpha-L-arabinofuranosides, and the total fermentation of AX has been associated with *Bifidobacterium longum* and *Bifidobacterium adolescentis* [8].

## 11.3   ARABINOXYLANS (AX) GELS

Gels are three-dimensional polymeric networks that absorb large amounts of water or biological fluids, granting the possibility of swelling, making them insoluble by their covalent crosslink [68].

## 11.3.1  OXIDATIVE CROSS-LINKING

AX has the ability to form covalent gels by FA coupling, resulting in the formation of di-AF and tri-AF [14]. AX gelling capacity is due to the AX structural characteristics such as Mw, A/X ratio, and FA content [45]. The formation process of the AX gels requires FA-oxidizing agents that act as crosslinking agents. These oxidizing agents can be chemical (ferric chloride or ammonium persulfate) and enzymatic (peroxidase/$H_2O_2$ or laccase/$O_2$) [7, 14] (Figure 11.2).

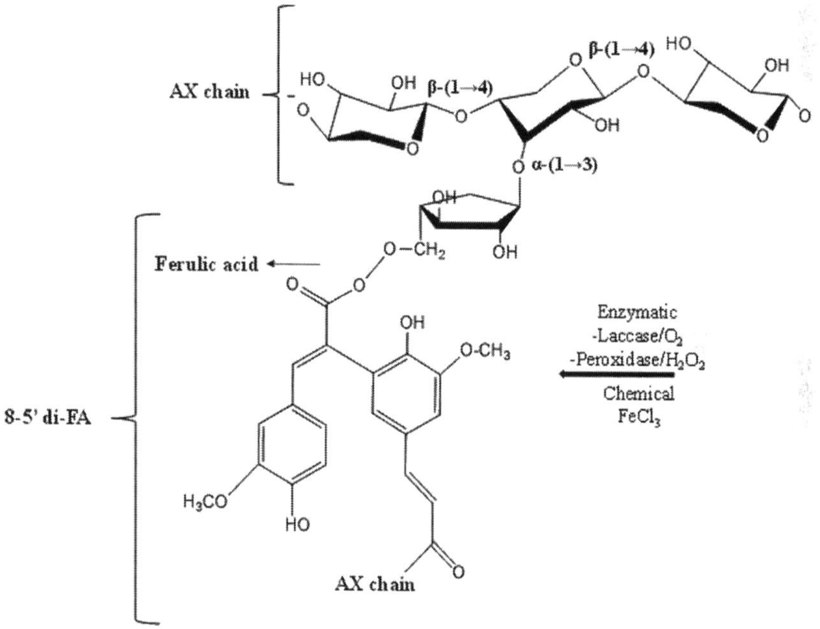

FIGURE 11.2   Schematic representation of gelation of feruloylated arabinoxylans. It is showing as an example of the possible formation of 8-5' di-FA.
*Source:* Adapted from: Niño-Medina et al. [14]; Mendez-Encinas et al. [71].

## 11.3.2  GELS CHARACTERISTICS

Carvajal-Millan et al. [39] formed gels in the presence of laccase with solutions of maize bran at 1 and 2% (w/v). These gels had a firmness of 0.32 and 0.81 N, respectively. The characteristics of the gels were related to the FA content; because with a higher FA content in the polysaccharide, the final gel structure is usually more compact due to its greater covalent crosslinking [45]. This type of gels can have various applications, e.g., the delivery of therapeutic proteins to the colon due to the mesh and macroporous structure (mesh size between 100 and 400 nm) [45, 69]. The stability of AX gels to changes in temperature, pH, and ionic strength would allow their passage through the upper gastrointestinal tract (GIT) and reach the colon. Here, AX can be fermented by gut microbiota. Therefore, AX gels could have potential applications for colon-specific biomolecules [5, 14] or cells such as probiotics [70]. AX gels have particular characteristics such as neutral taste or odor, capable of retaining water, stable to changes in pH, temperature, and ionic strength, all due to their covalent crosslinks [4, 45]. AX gel's capacity to absorb water gives them potential applications such as food additives, enzyme immobilizers, and controlled release devices [69, 72] and in the medical and pharmaceutical industries [68].

### 11.3.3   MACROGELS, MICROGELS, AND NANOGELS

Materials such as macrogels, microgels, and nanogels are formed when chemically or physically crosslinked synthetic or natural monomers are polymerized. In the formation of microspheres, macro, and microgels morphologies are observed. These materials have been widely studied due to their diverse applications in scientific and industrial sectors [19, 73–75]. Macrogels, also called macroporous gels, unlike micro and nanoparticles, are materials with pores greater than 50 nm, while the microgels are gel particles with sizes ranging from 0.1–100 μm. The term nanogel, according to IUPAC, is defined as a structure with a network size between 1 and 100 nm [19].

Previous investigations have reported various materials based on AX gels such as hydrogels, aerogels, films, beads, microspheres, and nanoparticles, which have been developed for different purposes (Figure 11.3). Microgels and nanogels are polymeric materials with more significant potential for biomedical applications because they offer a large surface area, and their internal network can be used for the incorporation and retention of bioactive molecules [76]. The maize bran AX microspheres reported by Martínez-López et al. [7] registered an average diameter of 350 μm. These microspheres presented a heterogeneous microstructure, which was attributed to the content and distribution of structures covalently crosslinked by FA, allowing the formation of the network. AX microspheres have shown advantages over other microspheres made from different types of polysaccharides because the networks covalently crosslinked are stable to changes in pH. Astiazarán-Rascón et al. [15] reported the formation of nanoparticles based on maize bran AX, which presented a diameter of approximately 193 nm, spherical form, and did not show the formation of agglomerates (Table 11.1).

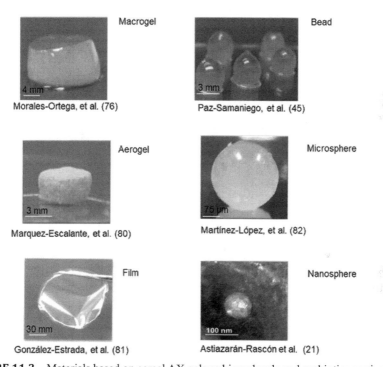

**FIGURE 11.3**   Materials based on cereal AX gels as biomolecule and probiotics carriers.

**TABLE 11.1**   Materials Based on Cereal AX Gels and Potential Application

| Material | Application | References |
|---|---|---|
| Macrogel | Probiotic entrapment | Morales-Ortega et al. [70] |
| Hydrogel | Caffeine entrapment | Iravani et al. [77] |
| Aerogel | – | Marquez-Escalante et al. [78] |
| Film | Biological control of postharvest diseases | González-Estrada et al. [79] |
| Film | Nanofibrillated cellulose composite | Stevanic et al. [80] |
| Pearls | Insulin and probiotics vehicle | Paz-Samaniego et al. [38] |
| Microsphere | Insulin delivery vehicle | Martínez-López et al. [81] |
| Nanosphere | Insulin delivery vehicle | Astiazarán-Rascón et al. [15] |

## 11.4  CONCLUSION

AX functionality derives from its chemical structure. Understanding this relationship enables the design of this polymer for a given application. AX has been successfully used to design macro, micro, and nanogels as delivery systems for several purposes. Besides, AX could be used for the reliable design of tailored gelling agents presenting concomitant bioactivity due to the FA content preserved within the molecule. Nevertheless, it is necessary to perform additional research focused on other AX gels characteristics such as prebiotic, antioxidant, and antiproliferative activity to evaluate this biomaterial not only as a simple carrier but also as a bioactive delivery system.

## KEYWORDS

- **arabinose/xylose**
- **arabinoxylans**
- **distillers dried grains with solubles**
- **ferulic acid**
- **molecular weight**
- **water-extractable**

## REFERENCES

1. Khafagy, E. S., Morishita, M., Onuki, Y., & Takayama, K., (2007). Current challenges in non-invasive insulin delivery systems: A comparative review. *Adv. Drug Deliv. Rev., 59*, 1521–1546. https://doi.org/10.1016/j.addr.2007.08.019.
2. Sarmento, B., Ribeiro, A., Veiga, F., Ferreira, D., & Neufeld, R., (2007). Oral bioavailability of insulin contained in polysaccharide nanoparticles. *Biomacromolecules, 8*, 3054–3060. https://doi.org/10.1021/bm0703923.
3. Carino, G. P., Jacob, J. S., & Mathiowitz, E., (2000). Nanosphere based oral insulin delivery. *J. Control. Release, 65*, 261–269. https://doi.org/10.1016/S0168-3659(99)00247-3.
4. Coviello, T., Matricardi, P., Marianecci, C., & Alhaique, F., (2007). Polysaccharide hydrogels for modified release formulations. *J. Control. Release, 119*, 5–24. https://doi.org/10.1016/j.jconrel.2007.01.004.
5. Carvajal-Millan, E., Guigliarelli, B., Belle, V., Rouau, X., & Micard, V., (2005). Storage stability of laccase induced arabinoxylan gels. *Carbohydr. Polym., 59*, 181–188. https://doi.org/10.1016/J.CARBPOL.2004.09.008.

6. Chen, M. C., Sonaje, K., Chen, K. J., & Sung, H. W., (2011). A review of the prospects for polymeric nanoparticle platforms in oral insulin delivery. *Biomaterials, 32*, 9826–9838. https://doi.org/10.1016/j.biomaterials.2011.08.087.

7. Martínez-López, A., Carvajal-Millan, E., Miki-Yoshida, M., Alvarez-Contreras, L., Rascón-Chu, A., Lizardi-Mendoza, J., & López-Franco, Y., (2013). Arabinoxylan microspheres: Structural and textural characteristics. *Molecules, 18*, 4640–4650. https://doi.org/10.3390/molecules18044640.

8. Martínez-López, A. L., Carvajal-Millan, E., Micard, V., Rascón-Chu, A., Brown-Bojorquez, F., Sotelo-Cruz, N., López-Franco, Y. L., & Lizardi-Mendoza, J., (2016). *In vitro* degradation of covalently cross-linked arabinoxylan hydrogels by bifidobacterial. *Carbohydr. Polym., 144*, 76–82. https://doi.org/10.1016/j.carbpol.2016.02.031.

9. Goldberg, M., Langer, R., & Jia, X., (2007). Nanostructured materials for applications in drug delivery and tissue engineering. *J. Biomater. Sci. Polym. Ed., 18*, 241–268. https://doi.org/10.1163/156856207779996931.

10. Sinha, V. R., & Kumria, R., (2001). Polysaccharides in colon-specific drug delivery. *Int. J. Pharm., 224,* 19–38. https://doi.org/10.1016/S0378-5173(01)00720-7.

11. Patel, M. M., (2013). Colon targeting: An emerging frontier for oral insulin delivery. *Expert Opin. Drug Deliv., 10*, 731–739. https://doi.org/10.1517/17425247.2013.7822 84.

12. Zhang, Z., Smith, C., & Li, W., (2014). Extraction and modification technology of arabinoxylans from cereal by-products: A critical review. *Food Res. Int., 65*, 423–436. https://doi.org/10.1016/j.foodres.2014.05.068.

13. Reis, S. F., Coelho, E., Coimbra, M. A., & Abu-Ghannam, N., (2015). Influence of grain particle sizes on the structure of arabinoxylans from brewer's spent grain. *Carbohydr. Polym., 130*, 222–226. https://doi.org/10.1016/j.carbpol.2015.05.031.

14. Niño-Medina, G., Carvajal-Millan, E., Rascon-Chu, A., Marquez-Escalante, J. A., Guerrero, V., & Salas-Muñoz, E., (2010). Feruloylated arabinoxylans and arabinoxylan gels: Structure, sources and applications. *Phytochem. Rev., 9*, 111–120. https://doi.org/10.1007/s11101-009-9147-3.

15. Astiazarán-Rascón, I., (2017). *Obtaining and Partial Characterization of Insulin-Loaded Arabinoxylan Nanoparticles.* Bachelor's thesis, University of Sonora. Hermosillo, Mexico.

16. Kumari, A., Yadav, S. K., & Yadav, S. C., (2010). Biodegradable polymeric nanoparticles based drug delivery systems. *Colloids Surfaces B Biointerfaces, 75*, 1–18. https://doi.org/10.1016/j.colsurfb.2009.09.001.

17. Hani, N., Azarian, M. H., Torkamani, A. E., & Kamil, M. W. A., (2016). Characterization of gelatin nanoparticles encapsulated with *Moringa oleifera* bioactive extract. *Int. J. Food Sci. Technol., 51*, 2327–2337. https://doi.org/10.1111/ijfs.13211.

18. Ávalos, A., Haza, A., & Morales, P., (2013). Silver nanoparticles: Applications and toxic risks to human health and environment. *Rev. Complut. Ciencias Vet., 7*, 1–23. https://doi.org/10.5209/rev_RCCV.2013.v7.n2.43408.

19. Kudaibergenov, S. E., Nuraje, N., & Khutoryanskiy, V. V., (2012). Amphoteric nano-, micro- and macrogels, membranes, and thin films. *Soft Matter., 8*, 9302. https://doi.org/10.1039/c2sm25766a.

20. Jung, T., Kamm, W., Breitenbach, A., Kaiserling, E., Xiao, J. X., & Kissel, T., (2000). Biodegradable nanoparticles for oral delivery of peptides: Is there a role for polymers to

affect mucosal uptake? *Eur. J. Pharm. Biopharm., 50*, 147–160. https://doi.org/10.1016/S0939-6411(00)00084-9.

21. Bhatia, S., (2016). Nanoparticles types, classification, characterization, fabrication methods, and drug delivery applications. In: *Nat. Polym. Drug Deliv. Syst.,* (pp. 33–93). Springer International Publishing, Cham. https://doi.org/10.1007/978-3-319-41129-3_2.

22. Soppimath, K. S., Aminabhavi, T. M., Kulkarni, A. R., & Rudzinski, W. E., (2001). Biodegradable polymeric nanoparticles as drug delivery devices. *J. Control. Release, 70*, 1–20. https://doi.org/10.1016/S0168-3659(00)00339-4.

23. Raies, A. B., & Bajic, V. B., (2016). *In silico* toxicology: Computational methods for the prediction of chemical toxicity, Wiley Interdiscip. *Rev. Comput. Mol. Sci., 6*, 147–172. https://doi.org/10.1002/wcms.1240.

24. Wani, M. Y., Hashim, M. A., Nabi, F., & Malik, M. A., (2011). Nanotoxicity: Dimensional and morphological concerns. *Adv. Phys. Chem.,* 1–15. https://doi.org/10.1155/2011/450912.

25. Buzea, C., Pacheco, I. I., & Robbie, K., (2007). Nanomaterials and nanoparticles: Sources and toxicity. *Biointerphases, 2*, MR17–MR71. https://doi.org/10.1116/1.2815690.

26. Dos, S. C. A., Seckler, M. M., Ingle, A. P., Gupta, I., Galdiero, S., Galdiero, M., Gade, A., & Rai, M., (2014). Silver nanoparticles: Therapeutical uses, toxicity, and safety issues. *J. Pharm. Sci., 103*, 1931–1944. https://doi.org/10.1002/jps.24001.

27. SCENIHR, (2015). Guidance on the determination of potential health effects of nanomaterials used in medical devices. *Sci. Comm. Emerg. New. Identified Heal. Risks,* 77. http://ec.europa.eu/health/scientific_committees/emerging/opinions/index_en.htm (accessed on 22 December 2020).

28. SCENIHR, (2009). Risk Assessment of Products of Nanotechnologies. *Sci. Comm. Emerg. New. Identified Heal. Risks,* 71. http://ec.europa.eu/health/scientific_committees/emerging/opinions/scenihr_opinions_en.htm (accessed on 22 December 2020).

29. Liu, Z., Jiao, Y., Wang, Y., Zhou, C., & Zhang, Z., (2008). Polysaccharides-based nanoparticles as drug delivery systems. *Adv. Drug Deliv. Rev., 60*, 1650–1662. https://doi.org/10.1016/j.addr.2008.09.001.

30. Lee, J. W., Park, J. H., & Robinson, J. R., (2000). Bioadhesive based dosage forms: The next generation. *J. Pharm. Sci., 89*, 850–866. https://doi.org/10.1002/1520-6017(200007)89:7<850::AID-JPS2>3.0.CO;2-G.

31. Carvajal-Millan, E., Landillon, V., Morel, M. H., Rouau, X., Doublier, J. L., & Micard, V., (2005). Arabinoxylan gels: Impact of the feruloylation degree on their structure and properties. *Biomacromolecules, 6*, 309–317. https://doi.org/10.1021/bm049629a.

32. Izydorczyk, M. S., & Biliaderis, C. G., (1995). Cereal arabinoxylans: Advances in structure and physicochemical properties. *Carbohydr. Polym., 28*, 33–48. https://doi.org/10.1016/0144-8617(95)00077-1.

33. Saulnier, L., Marot, C., Chanliaud, E., & Thibault, J. F., (1995). Cell wall polysaccharide interactions in maize bran. *Carbohydr. Polym., 26*, 279–287. https://doi.org/10.1016/0144-8617(95)00020-8.

34. Ruiz-Ruiz, J., Betancur-Ancona, D., González, R., & Chel-Guerrero, L., (2012). Extrusion of quality protein maize (*Zea mays* L.) in combination with hard-to-cook bean (*Phaseolus vulgaris* L.). In: *Maize Cultiv. Uses Heal. Benefits* (pp. 75–88). Nova Science Publishers, Inc. New York.

35. FIRA, (2016). *Trusts Established in Relation to Agriculture*. Maize agrifood panorama. https://www.gob.mx/cms/uploads/attachment/file/200637/Panorama_Agroalimentario_ Ma_z_2016.pdf (accessed on 22 December 2020).

36. Robles-Ramirez, M., Flores-Morales, A., & Mora-Escobedo, R., (2012). Maize tortillas: Physicochemical, structural, and functional changes. In: Jimenez-Lopez, J. C., (ed.), *Maize Cultiv. Uses Heal. Benefits* (pp. 89–111). Nova Science Publishers, Inc, New York.

37. Niño-Medina, G., Carvajal-Millan, E., Lizardi, J., Rascón-Chu, A., & Gardea, A. A., (2011). Feruloylated arabinoxylans recovered from low-value maize by-products. In: Jones, C. E., (ed.), *Handb. Carbohydr. Polym. Dev. Prop. Appl.* (pp. 711–725). Nova Science Publishers, New York, USA.

38. Paz-Samaniego, R., Carvajal-Millan, E., Sotelo-Cruz, N., Brown, F., Rascón-Chu, A., López-Franco, Y. L., & Lizardi-Mendoza, J., (2016). Maize processing wastewater upcycling in Mexico: Recovery of arabinoxylans for probiotic encapsulation. *Sustain, 8*, 1104. https://doi.org/10.3390/su8111104.

39. Carvajal-Millan, E., Rascón-Chu, A., Márquez-Escalante, J. A., Micard, V., De León, N. P., & Gardea, A., (2007). Maize bran gum: Extraction, characterization and functional properties. *Carbohydr. Polym., 69*, 280–285. https://doi.org/10.1016/j. carbpol.2006.10.006.

40. Li, S., Xiong, Q., Lai, X., Li, X., Wan, M., Zhang, J., Yan, Y., et al., (2016). Molecular modification of polysaccharides and resulting bioactivities. *Compr. Rev. Food Sci. Food Saf., 15*, 237–250. https://doi.org/10.1111/1541–4337.12161.

41. Martínez-López, A. L., Carvajal-Millan, E., Lizardi-Mendoza, J., López-Franco, Y., Rascón-Chu, A., & Salas-Muñoz, E., (2012). Ferulated arabinoxylans as by-product from maize wet-milling process: Characterization and gelling capability. In: Jimenez-Lopez, J. C., (ed.), *Maize Cultiv. Uses Heal. Benefits* (pp. 65–74). Granade, Spain.

42. Morales-Burgos, A. M., Carvajal-Millan, E., López-Franco, Y. L., Rascón-Chu, A., Lizardi-Mendoza, J., Sotelo-Cruz, N., Brown-Bojórquez, F., et al., (2017). Syneresis in gels of highly ferulated arabinoxylans: Characterization of covalent crosslinking, rheology, and microstructure. *Polymers (Basel), 9*, 164. https://doi.org/10.3390/ polym9050164.

43. Rodríguez, V., Revilla, P., & Ordás, B., (2012). New perspectives in maize breeding. In: Jimenez-Lopez, J. C., (ed.), *Maize Cultiv. Uses Heal. Benefits* (pp. 27–48). Nova Science Publishers, Inc, New York.

44. Mendez-Encinas, M. A., Carvajal-Millan, E., Yadav, M. P., López-Franco, Y. L., Rascon-Chu, A., & Lizardi-Mendoza, J., (2019). Partial removal of protein associated with arabinoxylans: Impact on the viscoelasticity, crosslinking content, and microstructure of the gels formed. *J. Appl. Polym. Sci., 136*. https://doi.org/10.1002/ app.47300.

45. Carvajal-Millan, E., Landillon, V., Morel, M. H., Rouau, X., Doublier, J. L., & Micard, V., (2005). Arabinoxylan gels impact of the feruloylation degree on their structure and properties. *Biomacromolecules, 6*, 309–317.

46. Ayala-Soto, F. E., Serna-Saldívar, S. O., García-Lara, S., & Pérez-Carrillo, E., (2014). Hydroxycinnamic acids, sugar composition, and antioxidant capacity of arabinoxylans extracted from different maize fiber sources. *Food Hydrocoll, 35*, 471–475. https://doi. org/10.1016/j.foodhyd.2013.07.004.

47. Vansteenkiste, E., Babot, C., Rouau, X., & Micard, V., (2004). Oxidative gelation of feruloylated arabinoxylan as affected by protein. Influence on protein enzymatic hydrolysis. *Food Hydrocoll., 18*, 557–564. https://doi.org/10.1016/j. foodhyd.2003.09.004.

48. Niño-Medina, G., Carvajal-Millán, E., Lizardi, J., Rascon-Chu, A., Marquez-Escalante, J. A., Gardea, A., Martinez-Lopez, A. L., & Guerrero, V., (2009). Maize processing wastewater arabinoxylans: Gelling capability and crosslinking content. *Food Chem., 115*, 1286–1290. https://doi.org/10.1016/j.foodchem.2009.01.046.

49. Biliaderis, C., & Izydorczyk, M., (2007). *Functional Food Carbohydrates*. Taylor & Francis Group, LLC, Boca Raton, FL. https://doi.org/10.1016/j.tifs.2007.11.004.

50. Saulnier, L., Guillon, F., Sado, P. E., Chateigner-Boutin, A. L., & Rouau, X., (2013). Plant cell wall polysaccharides in storage organs: Xylans (food applications). In: *Ref. Modul. Chem. Mol. Sci. Chem. Eng.* Elsevier. https://doi.org/10.1016/ B978-0-12-409547-2.01493-1.

51. Martínez-López, A. L., C. M. E., Lizardi-Mendoza, J., lópez-Franco, Y. L., R. C. A., Salas-Muñoz, E., & Ramírez-Wong, B., (2012). Ferulated arabinoxylans as by-product from maize wet-milling process: Characterization and gelling capability. In: *Maize Cultiv. Uses Heal. Benefits* (pp. 65–74). Nova Science Publishers, Inc., New York.

52. Morales-Ortega, A., Carvajal-Millan, E., López-Franco, Y., Rascón-Chu, A., Lizardi-Mendoza, J., Torres-Chavez, P., Campa-Mada, A., et al., (2013). Characterization of water extractable arabinoxylans from spring wheat flour: Rheological properties and microstructure. *Molecules, 18*, 8417–8428. https://doi.org/10.3390/molecules18078417.

53. Andrewartha, A. B., Phillips, K. A., & Stone, D. R., (1979). Biochemistry department, ~ La Trobe University, Bundoora, ~ Victoria 3083 (Australia). *Carbohydr. Res., 77*, 191–204.

54. Marcotuli, I., Hsieh, S. Y., Lahnstein, J., Yap, K., Burton, R. A., Blanco, A., Fincher, G. B., & Gadaleta, A., (2016). Structural variation and content of arabinoxylans in endosperm and bran of durum wheat (*Triticum turgidum* L.). *J. Agric. Food Chem., 64*, 2883–2892. https://doi.org/10.1021/acs.jafc.6b00103.

55. Marquez-Escalante, J. A., Carvajal-Millan, E., Yadav, M. P., Kale, M., Rascon-Chu, A., Gardea, A. A., Valenzuela-Soto, E. M., et al., (2018). Rheology and microstructure of gels based on wheat arabinoxylans enzymatically modified in arabinose to xylose ratio. *J. Sci. Food Agric., 98*, 914–922. https://doi.org/10.1002/jsfa.8537.

56. FDA, (2013). *GRAS Notices*. Food Drug Adm. Everything Added to Food United States (EAFUS). https://www.accessdata.fda.gov/scripts/fdcc/?set=GRASNotices&id=343& sort=Substance&order=ASC&startrow=1&type=basic&search=343 (accessed on 22 December 2020).

57. Aguedo, M., Fougnies, C., Dermience, M., & Richel, A., (2014). Extraction by three processes of arabinoxylans from wheat bran and characterization of the fractions obtained. *Carbohydr. Polym., 105*, 317–324. https://doi.org/10.1016/j.carbpol.2014.01.096.

58. Maes, C., & Delcour, J. A., (2002). Structural characterization of water-extractable and water-unextractable arabinoxylans in wheat bran. *J. Cereal Sci., 35*, 315–326. https:// doi.org/10.1006/jcrs.2001.0439.

59. Mendis, M., Leclerc, E., & Simsek, S., (2016). Arabinoxylans, gut microbiota and immunity. *Carbohydr. Polym., 139*, 159–166.

60. Mendis, M., & Simsek, S., (2014). Food Hydrocolloids Arabinoxylans and human health. *Food Hydrocoll., 42*, 239–243. https://doi.org/10.1016/j.foodhyd.2013.07.022.

61. Malunga, L. N., & Beta, T., (2016). Isolation and identification of feruloylated arabinoxylan mono- and oligosaccharides from undigested and digested maize and wheat. *Heliyon, 2*, e00106. https://doi.org/10.1016/j.heliyon.2016.e00106.

62. Zhang, P., Zhang, Q., & Whistler, R. L., (2003). L-arabinose release from arabinoxylan and arabinogalactan under potential gastric acidities. *Cereal Chem., 80*, 252–254. https://doi.org/10.1094/CCHEM.2003.80.3.252.

63. Wu, Y., & Zhang, G., (2018). Synbiotic encapsulation of probiotic *Lactobacillus plantarum* by alginate-arabinoxylan composite microspheres. *LWT-Food Sci. Technol., 93*, 135–141. https://doi.org/10.1016/j.lwt.2018.03.034.

64. Delzenne, N. M., Neyrinck, A. M., Bäckhed, F., & Cani, P. D., (2011). Targeting gut microbiota in obesity: Effects of prebiotics and probiotics. *Nat. Rev. Endocrinol., 7*, 639–646. https://doi.org/10.1038/nrendo.2011.126.

65. Sanders, M. E., (2008). Probiotics: Definition, sources, selection, and uses. *Clin. Infect. Dis., 46*, S58–S61. https://doi.org/10.1086/523341.

66. Kechagia, M., Basoulis, D., Konstantopoulou, S., Dimitriadi, D., Gyftopoulou, K., Skarmoutsou, N., & Fakiri, E. M., (2013). Health benefits of probiotics: A review. ISRN: *Nutr., 2013*, 1–7. https://doi.org/10.5402/2013/481651.

67. Marquez-Escalante, J., Carvajal-Millan, E., Lopez-Franco, Y. L., Valenzuela-Soto, E. M., & Rascon-Chu, A., (2018). Prebiotic effect of arabinoxylans and arabinoxylan-oligosaccharides and the relationship with good health promotion. *Cienciauat., 13*, 146–164.

68. Peppas, N. A., (2000). Hydrogels in pharmaceutical formulations. *Eur. J. Pharm. Biopharm., 50*, 27–46.

69. Berlanga-Reyes, C. M., Carvajal-Millan, E., Hicks, K. B., Yadav, M. P., Rascón-Chu, A., Lizardi-Mendoza, J., Toledo-Guillén, A. R., & Islas-Rubio, A. R., (2014). Protein/Arabinoxylans gels: Effect of mass ratio on the rheological, microstructural, and diffusional characteristics. *Int. J. Mol. Sci., 15*, 19106–19118. https://doi.org/10.3390/ijms151019106.

70. Morales-Ortega, A., Carvajal-Millan, E., Brown-Bojorquez, F., Rascón-Chu, A., Torres-Chavez, P., López-Franco, Y. L., Lizardi-Mendoza, J., et al., (2014). Entrapment of probiotics in water extractable arabinoxylan gels: Rheological and microstructural characterization. *Molecules, 19*, 3628–3637. https://doi.org/10.3390/molecules19033628.

71. Mendez-Encinas, M. A., Carvajal-Millan, E., Rascon-Chu, A., Astiazaran-Garcia, H. F., & Valencia-Rivera, D. E., (2018). Ferulated arabinoxylans and their gels: Functional properties and potential application as antioxidant and anticancer agent. *Oxid. Med. Cell. Longev.*, 1–22. https://doi.org/10.1155/2018/2314759.

72. Hernández-Espinoza, A. B., Piñón-Muñiz, M. I., Rascón-Chu, A., Santana-Rodríguez, V. M., & Carvajal-Millan, E., (2012). Lycopene/arabinoxylan gels: Rheological and controlled release characteristics. *Molecules, 17*, 2428–2436. https://doi.org/10.3390/molecules17032428.

73. Burchard, W., (2007). Macrogels, microgels and reversible gels-What is the difference? *Polym. Bull., 58*, 3–14. https://doi.org/10.1007/s00289-006-0588-1.

74. Downey, J. S., (2001). Poly(divinylbenzene) microspheres as an intermediate morphology in crosslinking precipitation polymerization. *Macromolecules, 34*, 4534–4541.

75. Frank, R. S., Downey, J. S., Yu, K., & Stöver, D. H. H., (2002). Poly(divinylbenzene-alt-maleic anhydride) microgels: Intermediates to microspheres and macrogels

in crosslinking copolymerization. *Macromolecules, 35*, 2728–2735. https://doi. org/10.1021/ma001927m.

76. Cruz, A., García-Uriostegui, L., Ortega, A., Isoshima, T., & Burillo, G., (2017). Radiation grafting of N-vinylcaprolactam onto nano and macrogels of chitosan: Synthesis and characterization. *Carbohydr. Polym., 155*, 303–312. https://doi.org/10.1016/j. carbpol.2016.08.083.

77. Iravani, S., Fitchett, C. S., & Georget, D. M. R., (2011). Physical characterization of arabinoxylan powder and its hydrogel containing a methylxanthine. *Carbohydr. Polym., 85*, 201–207. https://doi.org/10.1016/j.carbpol.2011.02.017.

78. Marquez-Escalante, J., Carvajal-Millan, E., Miki-Yoshida, M., Alvarez-Contreras, L., Toledo-Guillén, A. R., Lizardi-Mendoza, J., & Rascón-Chu, A., (2013). Water extractable arabinoxylan aerogels prepared by supercritical $CO_2$ drying. *Molecules, 18*, 5531–5542. https://doi.org/10.3390/molecules18055531.

79. González-Estrada, R., Calderón-Santoyo, M., Carvajal-Millan, E., Jesús, A. V. F. D., Ragazzo-Sánchez, J. A., Brown-Bojorquez, F., & Rascón-Chu, A., (2015). Covalently cross-linked arabinoxylans films for *Debaryomyces hansenii* entrapment. *Molecules, 20*, 11373–11386. https://doi.org/10.3390/molecules200611373.

80. Stevanic, J. S., Bergström, E. M., Gatenholm, P., Berglund, L., & Salmén, L., (2012). Arabinoxylan/nanofibrillated cellulose composite films. *J. Mater. Sci., 47*, 6724–6732. https://doi.org/10.1007/s10853-012-6615-8.

81. Martínez-López, A. L., Carvajal-Millan, E., Sotelo-Cruz, N., Micard, V., Rascon-Chu, A., Prakash, S., Lizardi-Mendoza, J., et al., (2016). *Biodegradable Covalent Matrices for Insulin Delivery by Oral Route Directed to the Colon Activated by the Microbiota and Process for its Obtaining.* International Patent Application PCT/MX2016/000170.

# PART 3

# Food Waste Management

# CHAPTER 12

# Food Waste Management: A Case Study of the Employment of Artificial Intelligence for Sorting Technique

A. JAMALI and Z. F. A. MISMAN

*Department of Mechanical and Manufacturing Engineering, Faculty of Engineering, Jalan Datuk Mohd. Musa – 94300, University Malaysia Sarawak, Malaysia, E-mail: jannisa@unimas.my (A. Jamali)*

## ABSTRACT

Food waste (FW) is a subcategory of municipal solid waste (MSW). FW management has received growing interest from a national, regional, and international level due to social, economic, and environmental impacts. The situation in Malaysia is no exception. Sorting waste is a labor-intensive process, and the development of the sorting system plays a vital role. Several technologies have been reported for waste sorting systems, and many have successfully been implemented in the actual field. However, most of the work limited to sorting certain groups of materials. Recently, the field of Artificial Intelligence (AI) has drawn researchers' attention in various fields, including waste sorting. Thus, the purpose of this chapter is to gain an insight of employment of AI in waste sorting and how it can benefit the FW sorting. The research focuses on the most salient progress of the available method or technique in AI for waste sorting. Three different classifications of studies in sorting waste that make use of AI technology are assessed. It is followed by a presentation of challenges of artificial intelligent sorting techniques, suggesting that whether every single methodology can effectively provide the accuracy of sorting, flexibility, and systems' reliability. Some of the potential concerns and recommendations for future research are presented in the conclusion.

## 12.1  INTRODUCTION

OECD described waste as *"materials that are not prime products; that is, products produced for the market) for which the generator has no further use in terms of his/her purposes of production, transformation or consumption and of which he/she wants to dispose of."* Waste belongs into several types, including (a) wastewater such as sewerage, organic liquids, and wash water; (b) solid waste includes plastic, paper, and card, tins, and metals and ceramics and glass; (c) organic waste are used to make compost for agriculture; (d) recyclable waste includes all waste items that can be recycled and reuse; and (e) hazardous waste includes all type of rubbish that is flammable, toxic, corrosive, and reactive.

The World Bank reported the world generates about 2.01 billion tons of municipal solid waste (MSW), with at least 33% of the waste is poorly managed, and the number is foreseen to grow to 3.40 billion tons by 2050. In 2016, Global Recycling proclaimed that Malaysia produced approximately 38.2 thousand tons of waste, with 17.5% of the recycling rate. However, due to unseparated waste, more than 30% of recyclable materials are directly disposed to landfills hence resulting in high consumption of landfills.

Waste sorting, a process by which waste is segregated into a different element, is the simplest yet effective way to lessen the volume of waste dumped into landfills. Waste sorting is beneficial in many ways. It will significantly lessen the volume of waste thrown into landfills [1], reduce environmental pollution [2], and reduce the potential hazards [3]. Recycling and composting also become possible through waste sorting.

The research on waste management has drawn significant interest over the past three decades. This is demonstrated through the extensive survey papers carried out beforehand [4–6]. Most of the surveys carried out showed that the sorting activities focus on the MSW. In the most recent survey performed by Sathish et al. [7], the research emphasizes on the current technologies available to sort MSW for recycling purposes. There are various waste sorting technologies being utilized, including waste screening, air separation, ballistic separation, film grabber, magnetic separation, eddy current separation, and sensor technology. Those technologies are effective but limited for detecting one type or a certain type of waste according to its specialty. For example, an eddy current separator is used to only separate metal from other types of waste.

None of the surveys directly discusses on sorting food waste (FW). However, throughout the activities, it is considered that the system indirectly sorting the FW from solid waste. Sorting FW is a real challenge because of the immense range of subcategories in MSW. In MSW themselves have more than 20 subcategories to be considered, which makes the problems of such a system incredibly difficult and complicated [8]. A sensible classification of the waste is crucial to produce an appropriate sorting system. Intelligent waste sorting is believed to be the most efficient method in increasing the recycling rate and improve the productivity of waste sorting. It also helps in reducing the operational costs of the recycling plant. In this approach, a human is only required in supervising the sorting process; hence, this will reduce the health risks of the workers.

This chapter provides a thorough overview of intelligent sorting technologies. The aim of this chapter is to assist developers in identifying an effective strategy for improving FW sorting technology. The chapter also provides a brief evaluation of the various methods of sorting materials. The chapter also addresses in depth a variety of sorted waste materials. This study also addresses open research problems and provides recommendations for potential work in the area of intelligent waste sorting.

The chapter discusses the detail on the topic listed below:

1. This chapter elaborates on the state of art of artificial intelligent sorting techniques for various waste segments like metals, plastic, paper, and glass.
2. This chapter evaluates the three main artificial intelligent sorting techniques in terms of diversity of materials, the accuracy of sorting, flexibility, and reliability.

## 12.2   WASTE SORTING TECHNOLOGIES

Even though many recycling centers have implemented techniques such as a magnet to pull out metals and air filters to separate paper from heavier plastic but sorting via hand is till done to date. Not only that it is a dirty job, manually sorting waste could pose a threat to health. The adaption of the intelligent sorting system in the recycling industry is believed to be able to deal with the catastrophe. Furthermore, it will significantly reduce the operating cost of recycling centers, improving productivity and helping in reducing the number of potentially recyclable waste dumps into landfills.

Prior to the system development stage, the framework of the waste management system must be clear. However, waste management practices completely vary across regions, countries, and even within the country [9].

An example of a framework proposed in 2015 [10] suggested that there are three main waste categories that are organic wastes, recycling waste and non-recyclable wastes. FW falls under organic wastes. However, MSW contained a mixture of all these wastes. In Malaysia, almost two-third of MSW is FW [11]. Thus, the discussion of intelligent sorting technologies which comprise of MSW management indirectly discuss on FW management. The intelligence sorting technologies can be divided into three primary approaches which are:

- Single sensor data or multiple-sensor data fusion;
- Data fusion and/or machine learning; and
- Image detection and image classification using deep learning.

In the next section, the details of each category are explained by means of examples.

### 12.2.1   SINGLE SENSOR DATA/MULTIPLE SENSORS DATA FUSION

1. **Two Pressure Capacitive Sensors [12]:** ROCycle, a trash-sorting robot is developed by researches from Massachusets International Technology (MIT). ROCycle identifies the type of trash by touch. It is built with two pressure capacitive sensors in its' two grippers to successfully distinguish the paper, plastic, and metal waste. Since the sensors are conductive, the robot can identify metal objects rather quickly. The specific object is classified based on two parameters that are pressure value and size of the objects. ROCycle has accurately classified 27 objects with 87% accuracy in a mock recycling plant. Besides, further experiment shows that the robot can provide 78% accuracy to differentiate between "hard" and "soft" objects.

2. **Three Sensors [13]:** An Inductive and a capacitive proximity sensor and a photoelectric sensor. The proposed system comprise of transmitting and collection process which is able to sort solid waste such as wood, glass, plastic, and metal. The material is detected via different combination of sensor outputs since none of the single sensor able to identify the four materials individually. Besides, the range adjustment of the capacitive proximity sensor is manipulated in order to distinguish between plastic and glass. The system is control based on the unique combination of sensors output. The truth table is developed from this data, and this implies the intelligent decision making

of the system. The verification of unique combinations is tested, and sensor outputs agreed with the developed truth table. The experiment is carried out repeatedly with objects of different materials and sizes showed the sorting results are consistent.

3. **IR Sensor, Voltage Sensor, and Inductive Sensor [14]:** In this research, a sorting system with the robotic arm is developed. The recognition component utilizes three sensors to detect and differentiate between three different waste materials that are metal, organic, and plastic. At the beginning of the sorting process, the presence of waste object is recognized by IR sensor. The metallic objects are sorted using the inductive sensor. Meanwhile, a voltage sensor is utilized to sort out between plastic and organic objects. Plastic is a non-conductive element that acts like a resistance. Thus, the voltage is zero because no voltage will be passed through it. Unlike organic objects, there are conductive in nature, thus resulting in non-zero voltage value. The fusion of those output sensors is used for controlling the sorting purposes. From the experiment carried out, the results show 81.8% accuracy.

4. **Ultrasonic Sensors [15]:** This study develops an indoor trash detection and collection robot known as Autrebot. The robot is equipped with the specific algorithm to allow the autonomous exploration. The algorithm is developed from the inputs that are obtained from two ultrasonic sensors. Ultrasonic sensors measure the distance of the objects. Thus, the location of the object can be determined from the sensor outputs. This information is then used as an intelligent decision making for the system. This is followed by pick-up and removal of trash procedure which is accomplished by pneumatic actuators install in the system. The developed algorithm is tested both in static and dynamic environments to emulate the actual scenario. Besides, the algorithm also being assessed in multiple scenarios such as the best, worst, and general cases individually. Thus, performance can describe more reliable results. The outcome from the tests confirms that the new proposed algorithm can operate effectively for real-time applications.

5. **Optical Sensor [16]:** This chapter introduces a mechanical sorting system which employs an indirect sorting process by using an optical sensor. The system consists of a conveyor belt, data acquisition system, and mechanical device. The system is able to capture various parameters from the sensor input such as particle sizes and positions,

colors, and shapes of each waste particle. Thus, the sorting criterion of the system manipulates those parameters. In this work, a multi-parameters acquisition is required to recognize single waste particles. The 3D visual image is scanned using the triangulation principle. The captured image is processed and analyzed using LabView software. Once the waste has been identified, the compressed air nozzle, which is placed at the end of the conveyor belt, will sort the particles accordingly. The system validation is carried out using five categories of waste mixtures inclusive of nonferrous metal mixtures such as contained copper, brass, and aluminum particles, two types of polymer mixture with diverse color that is polypropylene (PP) and acrylonitrile butadiene styrene (ABS) and a mixture of euro coins and bottle covers. The results from the experiment show a promising sorting outcome. The findings on sorting rates for the aforementioned waste are ranging between 90–100%. This study presents a brand-new approach for multi-feature recognition of sensor-primarily based sorting technology.

### 12.2.2  DATA FUSION AND/OR MACHINE LEARNING

1. **3D Sensor and Machine Learning Algorithm [17]:** The ZenRobotics Recycler (ZRR) is developed to sort construction and demolition (CND) waste. The system comprises of three main components that is sensors, a control system, and industrial robots. The industrial robots are responsible to select and pick waste on a conveyor and place it into multiple chutes. It acts after receiving the inputs from the sensors. The action is control by the control structure. This control structure uses machine learning for object and material recognition. The key features of the machine learning problems are the classification objects' material and the grasp mechanism option for asymmetrical objects. The supervised learning type is used for material classification, which is a task-driven. Meanwhile, the grasping task employs a reinforcement learning whereby the algorithm learns to react to the environment. The performance of the system is enhanced by manually interpreting error in the system. The accumulated off-line data is exploited for the manual annotation. However, there are some cases whereby the system unable to recognize the materials

from the sensor images. The developed ZRR is already deployed to sort the CND waste, which proved its functionality and feasibility.

2. **Thermal Imaging and Otsu Thresholding [18, 19]:** The system uses a thermal imaging-based system for sorting recyclables (RB) waste from MSW samples. The focus of work is to classify metallic fractions (MFs) and non-metallic fractions (NMFs) of electronic waste using thermal imaging-based practice. The thermal imaging with long wave range (LWIR) of 8–15 μm is utilized in this system. The captured images are pre-processed and stored in an image database. The object' features are processed using Otsu's thresholding, and material's types are determined via the segmentation threshold values. Subsequently, the segmented thermograms are processed to extract features comprising of the mean intensity, standard deviation, and the image sharpness. The extracted features are then presented to the classifier model to recognize the type of recyclable group in the segmented thermograms. From the experiment, the method successfully classifies iron, plastic, paper, aluminum, stainless steel, and wood from MSW. The range of 85–96% accuracy is obtained for the results of the classification system. Those results are comparable with the current single-material classification techniques.

3. **Ultrasonic Sensor and Content-Based Image Retrieval (CBIR) Systems [20]:** The Recyclebot system is made from two waste-bins mounted on the robot at both sides to place the sorting waste. One bin is for recyclable materials and the other one is for non-recyclable materials. It consists of five different modules for navigation, image acquisition, image processing, and human-machine interface. The camera is used as image acquisition and placed at the center on the top of the system. Then, ZigBee mesh network is employed to transfer the image to the server. In image processing, soda cans, plastic bottles, and paper cups are among waste that being considered as recycle waste materials. The image features of those recyclable are pre-identified and stored in a database. The image processing is carried out in MATLAB software. However, the overall flow of execution is controlled by a Python driver. There is a communication engine between those two which allow them to connect. The system uses Content-Based Image Retrieval (CBIR) system to find images that are visually look alike to a query image. The intelligent decision-making is made by comparing the new image of the waste object with the data stored in a database. In this research, the image

features use to assess the similarity between the images is the color composition of the image. The MATLAB Computer Vision System Toolbox provides a customizable bag-of-features framework to implement an image retrieval system. Apart from that, the system uses the ultrasonic sensors to detect the presence of the object on the platform and updates this status to the drivetrain module. The system is tested to sort recyclable objects from general waste materials. The proposed image processing techniques give a good result but with a limited set of databases.

4.   **Hyperspectral Sensors and Fuzzy Spectral-Spatial Distributions Method [21]:** This nonferrous material contained in waste of electric and electronic equipment (WEEE) sorting system is developed using hyperspectral sensors. The chapter proposes redesigned classification scheme in the hyperspectral domain using fuzzy spectral-spatial distributions method. The main objective is to reduce the computational time so that the real-time sorting operation can be offered. Besides, another aim is to attain real-time operation while maintaining the classification accuracy results as comparable to the earlier operation. The proposed classification algorithm is divided into four stages that is hyperspectral data decorrelation, background removal and nonferrous particle labeling, data acquisition, and the calculation of spectral-spatial density (SSD) for each nonferrous particle and material classification. In the first stage, the data decorrelation is carried out using the unsupervised approach to unprocessed hyperspectral images based on spectral fuzzy sets. The process allows the characteristics associated with nonferrous materials to be extracted by groups of spectral bands rather than selective spectral bands. For the background segmentation and particle labeling, a sole decorrelated energy vector component is employed for robust background segmentation. The calculation of SSD also exploits the same fuzzy-based method analogous to the data decorrelation. Thus, the main novelty resides in the optimized computational approach that allows the real-time calculation of the SSDs in the context of nonferrous material sorting. The results from the experiments disclose that the system is able to classify the WEEE scarp with the accuracy of 96.87% at the rate of 2.28 m/s.

5.   **Millimeter-Wave (MMW) Imaging by the Radiometric [22]:** The chapter discusses an approach to improve the wastepaper sorting capability by using multiple sensors to provide complementary data.

The authors are implementing millimeter-wave (MMW) imaging by the radiometric method to classify the wastepaper material. The main advantage of using a radiometric MMW approach is the ability to measure the bulk properties of paper objects as optical, and IR radiation is highly attenuated by these materials and can only make a surface or near-surface measurements of material properties. The type of paper can be determined by measuring the reduction in intensity caused by transmission and scattering through the paper sample. Thermally emitted MMW radiation is attenuated when propagating through the sample layer. The electromagnetic characteristics of the material, transmittance, reflectance, absorption, and scattering due to surface roughness all influence the degree to which MMW radiation is attenuated by the sample. Using other sensor modalities, such as an optical band camera or IR camera that provide information, which is independent of the information obtained from MMW imager, allows a more definitive determination of the type of waste paper material present to be made. The experiment is carried out with several types of sheet paper with different densities. The result shows that high-density samples such as magazines and books can be separated from a less dense sample such as cardboard, newspaper, and thin booklets based on their differing loss value. However, this method is only confirmed to be reliable in sorting the wastepaper materials.

6. **Image-based Analysis Using Dot Grid [23]:** The research investigates the physical composition of mixed residual waste materials using the novel dot-grid approach. The proposed method demonstrates a basic technique to determine the composition of large quantities of waste materials. ERDAS Imagine software is used to assess the area covered by each component within the waste materials. The composition determined from the image analysis is compared with results from the physical hand sorting. The results show that there are differences of composition in both samples. The discrepancies are due to the variety of density values of the materials being investigated. Besides, there is evident that those components, such as film plastics and paper, are being over-estimated by the image analysis approach. Regardless of that, there is still a strong correlation between the two datasets with correlation ($r = 0.91$), with the lowest correlation being 0.55.

7. **Near-Infrared (NIR) Hyperspectral Technology and Complementary Troubleshooting (CT) Method [24]:** In this study, NIR

hyperspectral technology is used to identify construction waste. However, to deal with the complex working environment, the CT method is proposed to classify the construction waste via online. It uses different colors to represent the waste materials individually. The CT method utilized the advantages of the random forest (RF) algorithm and the extreme learning machine (ELM). RF provides better results for potential feature recognition. Meanwhile, ELM provides a higher rate for characteristic reflectivity recognition. This new method is able to complement each algorithm and eliminate unstable results. This work investigates the classification of six categories of construction waste by selecting common materials such as woods, plastics, bricks: red and blacks, concretes, and rubbers as identification objects. The training data consist of 150 samples of each waste material. The samples are collected from the waste landfill. All samples are left in their original form and stationed in one place for a certain timeframe to emulate the real waste environment. The accuracy of the proposed method can reach up to 100% in identifying 180 samples.

8.  **Principle Component Analysis-Support Vector Machine (PCA-SVM) [25]:** In this research, a machine vision-based system is made to separate nonferrous metals from the end of life vehicles (ELVs). The classification algorithm and operation parameters are optimized to improve the separation efficiency of the system. The classification algorithm is optimized using Principal component analysis (PCA) with three options of classifiers that is support vector machine (SVM), k-nearest neighbor (KNN), and decision tree. The training and testing results reveal that the principle component analysis/support vector machine (PCA-SVM) achieved the highest recognition accuracy that is 96.64%. Besides, the operation parameters are optimized using response surface methodology. It is evaluated base on the separation efficiency. The experimental results are obtained from fluent numerical simulation. It is found that the system obtains 88.80% separation accuracy and 89.85% separation purity. The results show that the system exhibits a considerable performance in the separation of nonferrous metals. 3D optical sensor and Fuzzy technique with Neural Network classifier [26]; this work proposed sorting technique for recyclable packaging in the unstructured work environment. The image on the conveyor belt is captured by a 3D optical sensor which comprised of a CCD camera and a laser beam.

The geometric features of the objects by using the fuzzy-based decision module. Then, the classification of waste materials is determined using a suitably trained multilayer perceptron (MLP) neural network. Various net structures have been analyzed to solve the classification problem. There are 295 elements use for training while 101 patterns use for validation purposes. The training is stopped at the minimum sum squared error (SSE) point. Then, the validation set is executed. The results show that the validation errors are within 2.97–3.97%, which indicates good performance. From the results obtained, Net 2 shows the best performance among all. Thus, it has been chosen to be implemented in the real cell.

9. **Hidden Markov Model (HMM) [27]:** In this work, waste management using an image-based system is presented. The system works as such that it detects the waste level of the waste bins and scheduling the bin collection by applying the Hidden Markov Model (HMM) into the system. The experiment carries out using three bins. The bins are being analyzed using an overhead camera. There are three steps in the architecture of the bin detection, classification, and scheduling system. First, the location of the three bins opening was detected and defined by using the Hough Transform. The waste level is then determined by utilizing the Four Law Masks and the SVM. Three classifiers were used to train the system, which is k-NN, MLP, and SVM and the findings of the classification were compared. The result shows that k-NN and MLP both recorded 93% and 98% in classifying the waste level while the SVM classifier achieved 99.73% of the classification rate. HMM is adopted to schedule the ideal bin collecting time in their system. The parameters of the HMM are computed by the Forward-Backward algorithm and improved by the Baum-Welch algorithm. It is then decrypted with the Vertebi algorithm. The system is trained using 100 training images and tested on 100 test images with each image contains three bins that are rectangular with an opening area.

10. **MLP based on the Feedforward Neural Network (FFNN) Model [28]:** The research presents image-based waste level bin detection. Three main frameworks are applied in this system, which are image acquisition, feature extraction. Hough Transform is adopted in this work in order to extract the feature of the training and testing images while the classification of the solid waste level content is done by applying MLP based on the feedforward neural network (FFNN)

model. There are 250 images have been used to train the system with 30 images is used as the training images and another 20 images of each class as testing images. The experimental result shows that for both waste solid-class and waste solid-grade achieved excellent performance with 98.75% and 82.93% in the classification of the waste level bin.

### 12.2.3   IMAGE DETECTION AND CLASSIFICATION USING DEEP LEARNING

1. **Deep Neural Network [29]:** A trash pickup robot has the ability to pick up, detect, and classify trash on the grass is proposed. The authors implement a deep neural network for garbage recognition in the robot's system so that able the robot to detect the garbage correctly. There are three main frameworks used to accomplish the goal, namely perception, object tracker, and navigation. The navigation system of the proposed robot is made possible by the application of SegNet and ResNet-34, deep neural network architecture, for ground segmentation and garbage detection. The recognition accuracy was tested with a dataset containing 40,000 training images and 7000 testing images in six classes, which are five-garbage class and one non-garbage class. The experimental results showed that the classification error for the test set was 8.12%, 9.89%, 9.06%, 14.32%, and 22.3% for the bottle, can, carton, plastic bag, and wastepaper category, respectively. The paper and plastic bag classification errors are greater compared to other categories due to the lack of visual features. The system achieved a deficient non-garbage classification error. This means that the system is capable of distinguishing between garbage and non-garbage objects with high accuracy. Meanwhile, the cleaning efficiency of the robot showed approximately the same results when two different pathways that is planning pathway and random pathway are conducted during the experiments. Apparently, a random pathway is easier to implement as there is no need for path planning to cover the entire cleaning area. The drawback of this robot is that it requires a phase of training in a simulated environment. The training phase will take much time. This robot also has the risk of failing in a new and real environment.

2. **Convolutional Neural Network (CNN)-Deep Learning [30]:** Following major smart city construction in China, the researcher suggests an approach for street cleaning evaluation. The projected approach is to implement mobile edge computation and deep learning. There are three key frameworks in this approach, namely data collection and feedback scheduling for local management, pre-processing data and modeling. The main objective of the data collection is to capture waste and street images. Two pieces of information are acquired at this stage; street image information and local management information. The cleaning vehicles equip with high-resolution cameras are used to collect street images to obtain the street image information. Specific waste collection trucks, which act as the mobile station, will take images on the street on a regular basis. Mobile edge processing is used to improve the efficiency of the system. The images obtained from the data collection will be the first input of the CNN network and the scale of the images has been modified to a suitable size. Mobile edge processing has also been used to filter out unnecessary images from the mobile station. The image data will be transmitted through the edge database for local management information. The administrator will schedule the cleaning staff to clean the reported area based on the data obtained from the mobile station. The faster region-convolutional neural network (R-CNN) has been used to identify images of the street trash. The trained R-CNN model will be used to detect the garbage on the street. The proposed system is being tested on the CAFFE platform in Nanjing, China. The database is self-established and labeled and classified into nine classes: paper, plastic bag, plastic bottle, peel, cigarette butts, cigarette case, leaves, cans, and waste. The labeling and classification are based on the VOC2007 database format. A total of 681 images data with a size of 400 × 420-pixels are collected and divided into three parts, 321 practice images, 260 test images, and 100 images for the validation set. From the test, 3,686s were used to process 8,000 street images compared to the cloud server that consumed 379,206s to process the same amount of street images. The Faster R-CNN classified obtains a recognition rate of 0.998 for leaves, 0.995 for cigarette butt, and 0.988 for leaves. Nevertheless, the model has yet to be tested on a rainy day and has only been tested on a sunny day, and the model will only be trained on common street garbage data.

3. **Another Research Implementing CNN Reported in Ref. [31]:** In this chapter, researchers introduce a new decision-making module, where the robotic system chooses how to sort the objects in an unsupervised way. The Unsupervised Robotic Sorting (URS) uses deep CNN feature extraction and standard clustering algorithms. An extensive experiment on various standard datasets to demonstrate the efficiency of the proposed image-clustering pipeline is accrued out. A complex real-world dataset containing images of objects under various background and lighting conditions are introduced to evaluate the robustness of our URS implementation. This dataset is used to fine-tune the design choices (CNN and clustering algorithm) for URS. Finally, a method combining our pipeline with ensemble clustering to use multiple images of each object is proposed.

4. **Deep Neural Network Using RecycleNet [32]:** The research proposes RecycleNet deep convolution neural network architecture to sort waste materials such as paper, glass, plastic metal, cardboard, and trash. The RecycleNet model altered the connection patterns of the skip connections inside dense blocks. Thus, RecycleNet is carefully optimized deep CNN architecture for classification of selected recyclable object classes. This novel model reduced the number of parameters in a 121-layered network from 7 million to about 3 million. Besides, Adam is used as optimizer. Adam is derived from adaptive moment estimation. The extension algorithm to stochastic gradient descent that has recently gained a wide range of use for deep learning applications in computer vision. Several different deep CNN architectures are investigated such as REssNet50, MobileNet, InceptionResNetV2, and DenseNet121 Xception with two different optimization methodologies. From the results, it shows that the new model is both faster and more flexible, which significantly give accurate results.

5. **Fast-RCNN [33]:** In this research, a robotic grasping system for sorting garbage based on machine vision is proposed. This system needs to identify and achieve the positioning of target objects in the complex background. Then, the manipulator is used to grab the sorting objects. This chapter uses the Fast-RCNN to classify objects using deep convolution network method to achieve the authenticity identification of target objects in complex backgrounds. Fast R-CNN is composed of two sub-nets that are region proposal generation (RPN) and VGG-16. Those two sub-nets are employed to obtain accurate

sorting activities which exclusively dedicated to attain object recognition and pose estimation. RPN shares full-image convolutional features with the detection network, which eventually predicts object bounds and abjectness scores at each position. The optimization work is carried out by modified VGG-16 net. The performance is improved when VGG16-net decouples the angle loss and original detection loss. In laboratory simulation, 1999 images are captured, and the target objects are bottles. The experiment of garbage sorting to identify bottles using the proposed algorithm provides a promising result. The manipulator is able to perform garbage sorting efficiently.

## 12.3  CHALLENGES OF ARTIFICIAL INTELLIGENT SORTING TECHNOLOGIES

A single sensor system is easy to develop. The control system is straightforward because the system is less complicated. Thus, the system offers low price. However, the system may have a huge downside in terms of reliability. This is because to identify the waste material from an unknown environment is a non-trivial task. Thus, single data is insufficient to provide a good result. Meanwhile, for the multi-sensor system, the system' accuracy can be upgraded. The system with multiple sensors is more flexible as compared to those implementing machine learning in their systems. This is because the system does not implement a complicated algorithm to train dataset, which is time consuming. It relies only on multiple sensors to supply additional data. This method is time-wise, however, as there is no need to train large datasets. Although the system can accurately identify specific waste, the system cannot detect other types of waste unless the various sensors are installed. As being mentioned earlier, MSW is defined to have 20 different subcategories of waste. If the system is to be implemented in Malaysia, this requires a wider spectrum for the system. It may not be practical to install several sensors in one system, which results in a complex system. The proper combination and integration of the system need to be carefully resolved. This can escalate the cost and create troublesome for future maintenance too.

The aim of utilizing data fusion and machine learning is to overcome the constraint of individual sensor or multi-sensor. This is believed to give a diversity of sorting materials. This technology also can improve the performance of the system by giving a clear and accurate classification in sorting activities. Thus, the system will produce reliable sorting estimation.

However, data correlation and data association make this option become challenging tasks. If this is not properly tackled, the improvement of the system may worsen the performance of the system.

Image detection and classification using deep learning provides a diversity of materials, the accuracy of sorting, and reliability. In the past, the major challenges can come across during various stages, which can be categorized by factors affecting image acquisition, processing, and analysis, image preprocessing, and image segmentation. Besides, for real-time detection, apart from object detection algorithms need to accurately classify the object, they need to be fast at prediction time to meet real-time demands. However, these challenges can be rectified with proper assessment methodologies. In the field of waste sorting, deep learning is yet to have profound exploration and still an open problem. Deep learning has a big potential in sorting MSW due to unlimited data can be manipulated.

A survey carried out in 2019 [34], discusses the digital readiness of digitalization waste management. The chapter suggested that the advancement of sensors in the market and the evolution of AI technology may accelerate the technology to the next level. Thus, it may benefit the FW management in nearly future.

## 12.4  DISCUSSION

The machine learning is considerably mature for the field of waste sorting. The research review on Machine learning algorithms suggested that SVM are less robust and less ideal for waste sorting, but it is relatively easy and straightforward to implement with a high degree of accuracy for small datasets. There are different types of waste and a large dataset comprising all types of waste required to train the model to classify and identify the waste accurately. Therefore, most waste sorting approaches were integrating a neural network in their system because the neural network performs better than conventional machine learning and could provide better levels of classification compared to SVMs. However, a large dataset is needed to train the model in order to accomplish the objective of the system. If the database trained in deep learning is too small, it may result in overfitting [35]. Overfitting refers to a model that fits in well with the training data. It normally happens if the model learns the detail and noise in the training data to the extent that it negatively impacts the model's performance on the latest data. It demonstrates that the model gathers and learns the noise or natural differences in training data as principles. The concern is that these principles do not apply to new data and have a negative effect on model generalization capability. It is, therefore, necessary to choose a suitable model based on datasets as it affects the performance of the model. Training neural network is time-consuming and much more tedious than using an off-the-shelf classifier such as SVM, yet it performs better when it applies to complex problems such as image classification, natural language processing and speech recognition.

Transfer learning is a popular technique in machine learning and deep learning as it helps to create accurate models in a timesaving way [36]. It allows the developers to begin from concepts that have been already learned to solve a problem rather than initiating the learning process from the beginning. In machine learning, transfer learning is typically demonstrated by the use of pre-trained models. A pre-trained model referred to as a model that has been trained to solve a similar problem with the one that the developer wants to solve on a broad dataset. Accordingly, the use of models from published literature such as VGG, Inception, and MobileNet is standard practice due to the computational cost of training of these models. The pre-trained model is normally re-used for its purposes by replacing the initial classifier and embedding a new classifier that suits the objective before finely tuning the model [37]. There are three approaches in order to fine-tune the model: first, by training the entire model. In this scenario, the architecture of the pre-trained model is trained according to the desired dataset. However, a large data set is needed because the model must learn from scratch. Secondly are by train some convolutional layers, while others are left frozen. In this case, the developers are playing with the duality by deciding the network weights they want to alter. Typically, when the dataset is low and has a large number of parameters, more layers are left frozen to prevent overfitting. By comparison, if the dataset is large and the number of parameters is small, the model can be improved by training more layers for the new task, as overfitting is not a problem. Finally, it is by freezing the convolutional base. This context refers to an extreme train circumstance or freezing trade-off. The fundamental concept is to retain the convolutional base in its initial form and utilize its outputs to feed the classifier. The pre-trained model is used as a fixed extraction method for features, which can be beneficial if developers lack computational power, the dataset is limited, or the pre-trained model solves a problem that is very close to their goals. For instance, Li et al. [38] combines SVM and CNN for object classifier and apply the third strategy. The authors use the neural network as the feature extraction while the SVM will perform the classification job. Meanwhile, research in 2018 [39] also develops the system by combining two neural networks to successfully built solid and liquid classifiers. The classifier achieved a high accuracy rate and uses a lesser amount of time compare to train each image compared to RCNN and Faster RCNN.

The waste sorting system must be robust and reliable to enhance the productivity of waste sorting activities. As has been discussed above, the robustness of a system hugely depends on the performance of the classifier. However, building the neural network from scratch is tedious and time-consuming; thus, there are research recommends on developing the networks by using transfer learning. Transfer learning introducing a timesaving method and would also prevent a problem such as overfitting or underfitting in the model from occurring. The accuracy of each model and time took for each model to train a dataset is discussed detailed in the literature review. It is essential to develop an accurate model, but it is also vital for a model to minimize the prediction speed. The parameters to tune the model could be the number of hidden layers, the number of neurons per layer, learning rates, and regularization parameters [40]. The time taken for the model to train a dataset depends on the machine used by the developers, and the type of model chose. Every model has a different number of convolutional layers; hence the number of the hidden layer and the neurons per layer are also different.

## 12.5  CONCLUSION

The chapter provides an insight of the artificial intelligent methodology of sorting waste. The waste sorting technology should be versatile and capable of sorting different types of waste through one system, rather than just sorting one category of waste. Many of the established systems integrated "intelligence" into the framework because it is the most effective approach for handling mixed waste. Among the three intelligence-sorting technologies, the machine learning method has been profoundly explored with various waste materials. The study of deep learning in this field is remaining open problems. The existence of deep learning is believed to offer a better platform for sorting FW in nearly future.

Future research seeking to use this methodology should attempt:

1.  To develop an effective intelligent MSW system using deep learning, which indirectly manages FW because Malaysia does not implement

waste segregation 100%; hence the chances of waste being mixed are high.

2. To develop a high accuracy of the pre-trained model for waste sorting using transfer learning as it will prevent the developers from establishing a system from scratch. Thus, a vast amount of time and helps to reduce the potential of overfitting occur in their model. This will enhance the productivity of the waste segregation industry.

## KEYWORDS

- **acrylonitrile butadiene styrene**
- **artificial intelligent**
- **complementary troubleshooting**
- **construction and demolition**
- **content-based image retrieval**
- **convolutional neural network**
- **end of life vehicles**
- **food waste**
- **sorting technique and municipal solid waste**

## REFERENCES

1. Giacomo, D. A., & Giuseppe, V. M., (2019). Recycling and waste generation: An estimate of the source reduction effect of recycling programs. *Ecological Economics, 161*, 321–329.

2. Relationship between Recycling Rate and Air Pollution, (2015). Waste management in the state of Massachusetts. *Waste Management, 40*, 192–203.

3. Ziraba, A. K., Haregu, T. N., & Mberu, B., (2016). A review and framework for understanding the potential impact of poor solid waste management on health in developing countries. *Archives of Public Health Archives Belges De Sante Publique, 74*, 55. https://doi.org/10.1186/s13690-016-0166-4.

4. Dodbiba, G., & Fujita, T., (2004). Progress in separating plastic materials for recycling. *Phys. Sep. Sci. Eng., 13*(3/4), 165–182.

5. Wu, G., Li, J., & Xu, Z., (2013). Triboelectrostatic separation for granular plastic waste recycling: A review. *Waste Manage, 33*(3), 585–597.

6. Rahman, M. O., Hussain, A., & Basri, H., (2014). A critical review on wastepaper sorting techniques. *Int. J. Environ. Sci. Technol., 11*(2), 551–564.

7. Gundupalli, S. P., Hait, S., & Thakur, A., (2017). A review on automated sorting of source-separated municipal solid waste for recycling. *Waste Manag., 60*, 56–74.

8. Wahidah, S., & Ghafar, A., (2017). *Food Waste in Malaysia: Trends, Current Practices and Key Challenges* (pp. 1–10).

9. Hoornweg, D., & Bhada-Tata, P., (2012). Urban development series, knowledge papers. Washington: World Bank. *What a Waste: A Global Review of Solid Waste Management.*

10. Elsaid, S., & Aghezzaf, E., (2015). A framework for sustainable waste management: Challenges and opportunities. *Management Research Review, 38*(10), 1086–1097.

11. Lim, W. J., Chin, N. L., Yusof, A. Y., Yahya, A., & Tee, T. P., (2016). Food waste handling in Malaysia and comparison with other Asian countries. *International Food Research Journal, 23*, S1–S6.

12. Conner-Simons, A., (2019). *Robots that Can Sort Recycling: CSAIL's "RoCycle" System Uses in-Hand Sensors to Detect if an Object is Paper, Metal or Plastic.* Retrieved from: http://news.mit.edu/2019/mit-robots-can-sort-recycling-0416 (accessed on 22 December 2020).

13. Chahine, K., & Ghazal, B., (2017). Automatic sorting of solid wastes using sensor fusion. *International Journal of Engineering and Technology, 9*(6), 4408–4414.

14. Diya, S. Z., et al., (2018). Developing an intelligent waste sorting system with robotic arm: A STEP towards green environment. In: *2018 International Conference on Innovation in Engineering and Technology (ICIET)* (pp. 1–6). Dhaka, Bangladesh.

15. Kulkarni, S., & Junghare, S., (2013). Robot-based indoor autonomous trash detection algorithm using ultrasonic sensors. In: *2013 International Conference on Control, Automation, Robotics, and Embedded Systems (CARE)* (pp. 1–5). Jabalpur.

16. Huang, J., Pretz, T., & Bian, Z., (2010). Intelligent solid waste processing using optical sensor-based sorting technology. In: *2010 3rd International Congress on Image and Signal Processing* (pp. 1657–1661). Yantai.

17. Lukka, T. J., Tossavainen, T., Kujala, J. V., & Raiko, T., (2014). ZenRobotics recycler-robotic sorting using machine learning. *Proceedings of the International Conference on Sensor-Based Sorting (SBS).*

18. Sathish, P. G., Subrata, H., & Atul, T., (2017). Multi-material classification of dry recyclables from municipal solid waste based on thermal imaging. *Waste Management, 70*, 13–21.

19. Sathish, P. G., Subrata, H., & Atul, T., (2018). Classification of metallic and non-metallic fractions of e-waste using thermal imaging-based technique. *Process Safety and Environmental Protection, 118*, 32–39.

20. Chinnathurai, B. M., Sivakumar, R., Sadagopan, S., & Conrad, J. M., (2016). *Design and Implementation of a Semi-Autonomous Waste Segregation Robot* (pp. 1–6). Southeast Con, Norfolk, VA.

21. Artzai, P., Aranzazu, B., Jone, E., Ovidiu, G., Paul, F. W., & Pedro, M. I., (2012). Real-time hyperspectral processing for automatic nonferrous material sorting. *J. Electron. Imag., 21*(1), 013018.

22. Guérin, J., Thiery, S., Nyiri, E., & Gibaru, O., (2018). Unsupervised robotic sorting: Towards autonomous decision-making robots. *Journal of Artificial Intelligence and Applications (IJAIA), 9*(2).

23. Wagland, S. T., Veltre, F., & Longhurst, P. J., (2012). Development of an image-based analysis method to determine the physical composition of a mixed waste material. *Waste Management, 32*(2), 245–248.

24. Wen, X., Jianhong, Y., Huaiying, F., Jiangteng, Z., & Yuedong, K., (2019). A robust classification algorithm for separation of construction waste using NIR hyperspectral system. *Waste Management, 90*, 1–9.

25. Chao, W., Zhili, H., Qiu, P., & Lin, H., (2019). Research on the classification algorithm and operation parameters optimization of the system for separating nonferrous metals from end-of-life vehicles based on machine vision. *Waste Management, 100*, 10–17.

26. Mattone, R., Campagiorni, G., & Galati, F., (2000). Sorting of items on a moving conveyor belt. Part 1: A technique for detecting and classifying objects. *Robotics and Computer-Integrated Manufacturing, 16*(2/3), 73–80.

27. Aziz, F., Arof, H., Mokhtar, N., Shah, N. M., Khairuddin, A. S. M., Hanafi, E., & Talip, M. S. A., (2018). Waste level detection and HMM-based collection scheduling of multiple bins. *PLoS One, 13*(8), 1–14. https://doi.org/10.1371/journal.pone.0202092.

28. Hannan, M. A., Zaila, W. A., Arebey, M., Begum, R. A., & Basri, H., (2014). Feature extraction using Hough transform for solid waste bin level detection and classification. *Environmental Monitoring and Assessment, 186*(9), 5381–5391. https://doi.org/10.1007/s10661-014-3786-6.

29. Bai, J., Lian, S., Liu, Z., Wang, K., & Liu, D., (2018). Deep learning-based robot for automatically picking up garbage on the grass. *IEEE Transactions on Consumer Electronics,* pp(c), 1. https://doi.org/10.1109/TCE.2018.2859629.

30. Zhang, P., Zhao, Q., Gao, J., Li, W., & Lu, J., (2019). Urban street cleanliness assessment using mobile edge computing and deep learning. *IEEE Access, 7*, 63550–63563. https://doi.org/10.1109/ACCESS.2019.2914270.

31. Shylo, S., & Harmer, S. W., (2016). Millimeter-wave imaging for recycled paper classification. *IEEE Sensors Journal, 16*(8), 2361–2366. https://doi.org/10.1109/JSEN.2015.2512106\[34].

32. Bircanoğlu, C., Atay, M., Beşer, F., Genç, Ö., & Kızrak, M. A., (2018). RecycleNet: Intelligent waste sorting using deep neural networks. In: *2018 Innovations in Intelligent Systems and Applications (INISTA)* (pp. 1–7). Thessaloniki.

33. Zhihong, C., Hebin, Z., Yanbo, W., Binyan, L., & Yu, L., (2017). A vision-based robotic grasping system using deep learning for garbage sorting. In: *2017 36th Chinese Control Conference (CCC)* (pp. 11223–11226). Dalian.

34. Sarc, R., Curtis, A., Kandlbauer, L., Khodier, K., Lorber, K. E., & Pomberger, R., (2019). Digitalization and intelligent robotics in value chain of circular economy-oriented waste management: A review. *Waste Management, 95*(7), 476–492.

35. Srivasta, N., Hinton, G., Krizhevsky, A., Sutskever, I., & Salakhutdinov, R., (2014). Dropout: A simple way to prevent neural networks from overfitting. *Journal of Machine Learning Research, 15*(1), 1929–1958.

36. Rawat, W., & Wang, Z., (2017). Deep convolutional neural networks for image classification: A comprehensive review. *Neural Computation, 29*(9), 2352–2449.

37. Van, E. J. E., & Hoos, H. H., (2020). A survey on semi-supervised learning. *Machine Learning, 109*(2), 373–440.

38. Li, N., Zhao, X., Yang, Y., & Zou, X., (2016). Objects classification by learning-based visual saliency model and convolutional neural network. *Computational Intelligence and Neuroscience, 2016*, 12.

39. Ramalingam, B., Lakshmanan, A. K., Ilyas, M., Le, A. V., & Elara, M. R., (2018). Cascaded machine-learning technique for debris classification in floor-cleaning robot application. *Applied Sciences (Switzerland), 8*(12), 1–19.

40. Michelucci, U., (2018). *Applied Deep Learning: A Case-Based Approach to Understanding Deep Neural Networks*. A Press.

# CHAPTER 13

# Food Wastes Management: Practice of Composting Process at the IIUM Kuantan Campus

MOHD. ARMI ABU SAMAH,[1] KAMARUZZAMAN YUNUS,[2]
DATO' WAN MOHD. HILMI WAN KAMAL,[3]
MOHD. RAMZI MOHD. HUSSAIN,[4] and RAZSERA HASSAN BASRI[3]

[1]*Assistant Professor, Kulliyyah of Science, IIUM Kuantan Campus, Malaysia*

[2]*Professor, Kulliyyah of Science, IIUM Kuantan Campus, Malaysia*

[3]*Development Division, IIUM Gombak, Malaysia*

[4]*Associate Professor, Kulliyyah of Architecture and Environmental Design, IIUM Gombak, Malaysia*

## ABSTRACT

Malaysian population has been increasing at a rate of 2.4% per year or about 600,000 per year since 1994. With this population growth, the municipal solid waste (MSW) generation also rises, which makes municipal solid waste management important. Changes in municipal solid waste generation rates are mostly caused by socioeconomic characteristics of a population and facilities, which are provided by the various departments. Malaysian solid waste contains a very high amount of organic waste and consequently has high moisture content and a bulk density above $200 kg/m^3$. Nowadays, the only way used for the managing and discarding of MSW in Malaysia is landfilling. The disposal of wastes through landfilling is becoming more complicated because vacant landfill sites are filling up at a very rapid rate. This article aims to shed light on the origins of food waste, both pre-consumer and post-consumer. Secondly, the impacts and challenges of cutting down on food waste produced in IIUM Kuantan Campus will also are discussed.

On top of that, this article will propose a new system which is by using a composting machine to manage food waste with proper management and cleans.

## 13.1  INTRODUCTION

The United Nations Food and Agricultural Organization (FAO) reported that, around the world, about 1.6 billion tons of food is lost and wasted every year. These food wastes (FWs) issues cause severe bad effects on the environmental and socioeconomic [1]. Bad management of FW gives rise to human health problems, the loss of natural resources, contamination of rivers and seas, the generation of methane emissions from landfill sand dumps, and the opportunity to recover valuable energy, nutrients, organic matter, and water contained in the FW are lessened. According to Morton et al. [2], all United Nation Member States as part of the United Nation's sustainable development goals (SDGs) committed to stop starvation, make sure easy access to hygienic water and sanitation for all, make cities sustainable (including by Global Food Waste Management: an Implementation Guide for Cities "paying special attention to municipal and other waste management") and take crucial action to fight climate change. Without the sustainable management of FW, this engagement cannot be met-so it is very important to take action on this front. As a result of continued global urbanization, 68% of the world's population will live in urban areas by 2050 [3], so there is a particularly pressing need to improve the management of FW in urban areas.

Other than that, the Malaysian population also has been increasing at a rate of 2.4% per year or about 600,000 per year since 1994. With this population growth, the municipal solid waste (MSW) generation also rises, which makes MSW management important. According to Kathirvale et al. [4], the average amount of MSW generated in Malaysia in 2003 was 0.5–0.8 kg/ person/day; it has rose to 1.7 kg/person/day in major cities. Thus, the amount of MSW generated was estimated to have risen to 31,000 tons by the year 2020. Information on the amount of solid waste generated is essential to almost all aspects of solid waste management (SWM) [5]. Most studies on MSW generation used the load-account analysis, which is based on waste collected and disposed in the landfills. Changes in MSW generation rates are mostly caused by socioeconomic characteristics of a population and facilities, which are provided by the various departments. Malaysian solid waste contains a very high amount of organic waste and consequently has high moisture content and a bulk density above 200 kg/m³. A waste characterization study found that the major items of Malaysian waste were food, plastic, and paper which comprise 80% of overall weight [4]. These characteristics reveal the natural world and daily life of the Malaysian population.

Nowadays, the only way used for the managing and discarding of MSW in Malaysia is landfilling. A majority of the landfill sites are open dumping spaces, which cause serious environmental and social pollution [6]. The disposal of wastes through landfilling is becoming more complicated because vacant landfill sites are filling up at a very rapid rate. At the same time, build up new landfill sites is becoming more complex because of land shortage and the rise in land prices and high demands, particularly in urban areas due to the addition amount in population. According to Wan and Kadir [7], there were 155 disposal sites under the responsibility of local authorities in Malaysia. These disposal sites ranging in size from 8 to 60 ha, as shown in Table 13.1, depending on the location and amount of waste disposed [8]. Most of these areas are open dumpsites, and the space has been an excessive amount of wastes. The operation of these sites has been extended due to the deficiency of suitable and cost-effective alternatives to treat the waste.

**TABLE 13.1**   Types and Number of Disposal Site in Malaysia

| State | Open Dumping | Controlled Dumping | Sanitary Landfill | Total |
|---|---|---|---|---|
| Johor | 12 | 14 | 1 | 27 |
| Kedah | 9 | 5 | 1 | 15 |
| Kelantan | 12 | 2 | 0 | 14 |
| Melaka | 2 | 3 | 0 | 5 |
| Negeri Sembilan | 8 | 6 | 0 | 14 |
| Pahang | 7 | 5 | 3 | 15 |
| Perak | 15 | 11 | 4 | 30 |
| Perlis | 0 | 1 | 0 | 1 |
| Pulau Pinang | 1 | 1 | 1 | 3 |
| Selangor | 5 | 15 | 0 | 20 |
| Terengganu | 2 | 8 | 1 | 11 |
| Total | 73 | 71 | 11 | 155 |

*Source:* Wan and Kadir [8].

At present, awareness of effective waste management already becomes important. The implementation of Act 672 required waste segregation makes the waste easier to undergo treatment. Recyclable waste will be collected and sold. Meanwhile, for organic waste and food waste (FW), it will be delivered to the compost site. The finished product from the FW composting is fertilizer. It can be used as a fertilizer and as a soil amendment. Islamic International University Malaysia (IIUM) Kuantan SWM a key result area (KRA) project that is developed to initiate a better system to manage solid waste in IIUM Kuantan Campus. It is a continuous daily basis operation that needs someone to supervise with a high commitment to guarantee that waste in the campus is properly handled in accordance with the underlined procedures. This project is a collaboration between IIUM Kuantan Campus management authorities, Daya Bersih Sdn. Bhd. (DBSB) IIUM Kuantan branch and cafeteria operators under the management of IIUM food and services unit (FSU). Basically, the objectives of this project are:

1. To treat waste generated at IIUM Kuantan Mahallah's cafeteria efficiently using a suitable waste treatment.
2. To encourage waste generator in IIUM Kuantan Campus to achieve zero-waste generation.
3. To build a special composting plant facility for organic waste treatment, especially FW at IIUM Kuantan Campus.

The project aims to shed light on the origins of FW, both pre-consumer and post-consumer. Secondly, the impacts and challenges of cutting down on FW produced in IIUM Kuantan Campus will also are discussed. On top of that, the project will propose a new system which is by using a composting machine to manage FW with proper management and cleans.

## 13.2    FOOD GENERATIONS IN IIUM KUANTAN CAMPUS

The FW generation by customers at all Mahallah's cafeteria, Kulliyah's cafeteria, and food operators around the IIUM Kuantan Campus such as cafeteria in One-Stop Student Center and cafeteria in Office of the Campus Director. Basically, the food generator in IIUM Kuantan Campus generates FW with many types such as:

- Raw or cooked fruit and vegetables;
- Raw or cooked meat and fish, including bones and skin;
- All non-liquid dairy products;
- Cakes, breads, and pastries;

- Uneaten food from customers' plates and dishes;
- Eggs including shells;
- Pasta, rice, and beans;
- Coffee grounds and tea bags.

All of these FWs will be thrown into the designated bin, namely is a smart bin to separate the food and its leachate. After that the FWs will be delivered to the composting site for composting process activity. However, the leachate water should be reused for landscaping and cultivated plants, and it must not discharge into a drain to avoid water and odor pollution.

## 13.3  MANAGEMENT OF FOOD WASTE (FW) COMPOSTING PROCESS IN IIUM KUANTAN CAMPUS

The waste generated by IIUM Kuantan Campus will be segregated at the source, meaning the FW was segregate at the Mahallahs' cafeteria. The recyclable waste will be sending or sold to the recycling center. Meanwhile, non-recyclable waste will be sent to a composting center under the special facility. Therefore, this compost facility will be equipped with the compost bin that can treat FW and organic wastes from the selected area in the campus. The finished product will be produced as organic fertilizer and will be used in IIUM Kuantan Campus area and the rest will be vended to the outsiders.

The flow chart of the composting project was shown in Figure 13.1. For the project management, it has five steps which are initiation, planning, execution, monitoring, and evaluation. For execution steps, the step to compost system was proposed. The waste was segregate first at the source of generation. This will be done by the Mahallahs' cafeteria owner. After that, the FW will be transport to the proposed compost plant. At the compost plant, the composting process will be done.

In a compost process, the main steps are drying, grinding, starting, and finished the process. When the FW arrived at the compost plant, it will be spread onto canvas to dry and get rid of excess water. After that, the FW will undergo a grinding process. This grinding method will accelerate the composting process by increasing the surface of contact with the microbe. With the addition of aeration, FW will decompose faster.

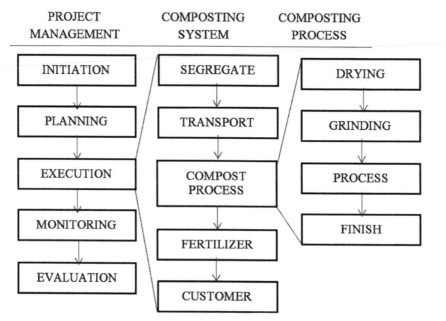

**FIGURE 13.1**   The flow chart diagram of the IIUM Kuantan Campus Composting Project.

Fundamentally, Figure 13.2 shows the key elements in SWM procedures in IIUM Kuantan. This figure will explain regarding flow processing starting from generation of waste from all sources until to product of FW organic fertilizer. Basically, all the food operators will be segregated their wastes into three categories which are recyclable waste, non-recyclable wastes and FW. The recyclable waste will be segregated followed their types such as papers, plastics, aluminum, and glass.

Meanwhile, food or organic waste will be thrown into a designated smart bin. Then all the recyclable waste will be stored in recyclable storage facilities while FWs will be stored in waste storage facilities. After the recyclable wastes and FWs have reached the limitation of the standard amount, the wastes will be collected and transported to the particular area. For FW, it will be transported to a composting area for the composting process, which will produce organic compost, while the recyclable waste will be transported to recycling site (Figure 13.2). Therefore, Figure 13.3 shows the flow chart diagram of the SWM system in IIUM Kuantan.

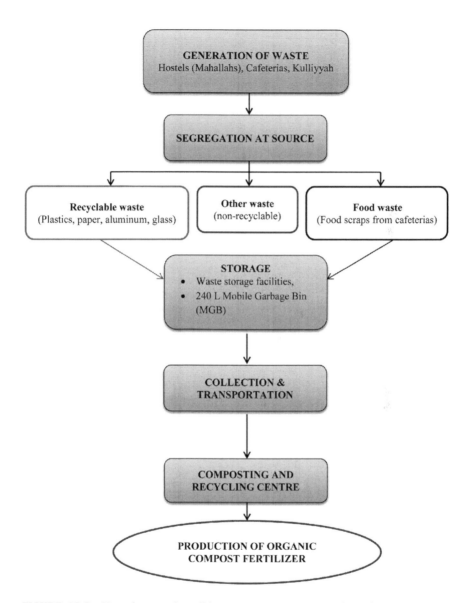

**FIGURE 13.2** Key elements in solid waste management procedures in IIUM Kuantan Campus.

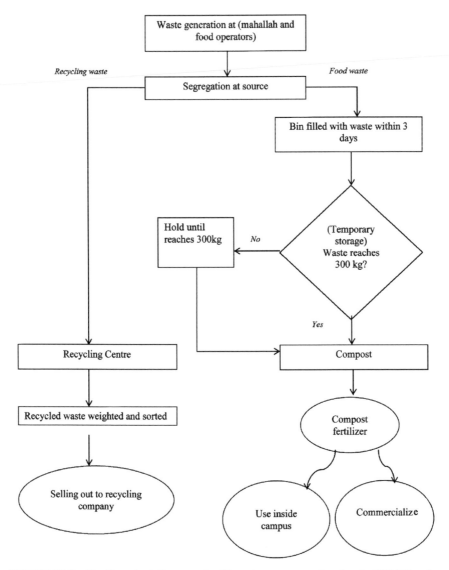

**FIGURE 13.3**  The flow chart diagram of solid waste management system in IIUM Kuantan Campus.

## 13.3.1  FOOD WASTE (FW) COMPOSTING PROCESS

FW that is not composted generally ends up in landfills. Degeneration process of FW produces a large amount of methane-a more vigorous greenhouse gas than even carbon dioxide which can cause global warming and climate change. So to reduce the amount of FW being deposited into landfills, DBSB in collaboration with IIUM Kuantan Campus management authorities and cafeteria operators under the management of IIUM FSU designated a FW composting project (Figure 13.4). The FW composting project is about a system that will treat FW and turn it into organic compost. The project is to encourage the best practice in SWM and also to implement acceptable guidelines on waste management procedures in IIUM Kuantan Campus to food generators and waste management operators. These guidelines introduce a standard code of practice for SWM.

**FIGURE 13.4**    Food waste composting project area.

Waste in IIUM Kuantan Campus was managed by DBSB. The waste generated was segregate at the source. These wastes were collected every day by DBSB workers (Figure 13.5). The FW will be sent to the compost plant (Figure 13.6) and the recyclable waste will be sold to the recycling center.

**FIGURE 13.5**    The collection of wastes by DBSB.

**FIGURE 13.6**    Composting plant.

The FWs will be collected and stored in the designated bin for a few days. The bin was built manually. It functions to separate the food and its leachate. Figure 13.7 shows the bin used for the collection of FW. The process for producing fertilizer normally takes 2 days to 4 days to complete. When the organic wastes reach at the compost site, the leachate from the bin will be transferred into a container. After that, the organic wastes will be weighed and transferred into a container. Assured amounts of dried leaves will be added into the organic wastes based on the ratio calculated (Figure 13.8), and the bacteria will be sprayed onto the mixture. The mixture was then stirred and ground before going into the compost machine. Figure 13.9 shows FW through a grinding process, and Figure 13.10 shows FWs are transferred into a rotary bin machine process. The compost machine will be rotated for 2 days (Figure 13.11). After 2 days, the mixture or known as fertilizer (Figure 13.12) will be out and spread out in a container to let it cool. Finally, the fertilizer is ready to be used. Figure 13.13 shows a basic flow of the compost process.

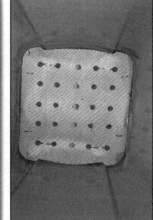

**FIGURE 13.7**   Smart bin facility.

**FIGURE 13.8**   Food waste ready to be mixed.

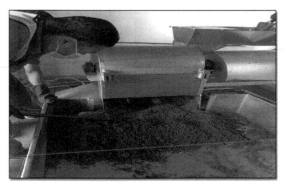

**FIGURE 13.9**   Food waste through a grinding process.

**FIGURE 13.10**   Food wastes are transferred into the rotary bin machine process.

**FIGURE 13.11**   Compost machine.

**FIGURE 13.12**   Food wastes composting process end product and ready to be used.

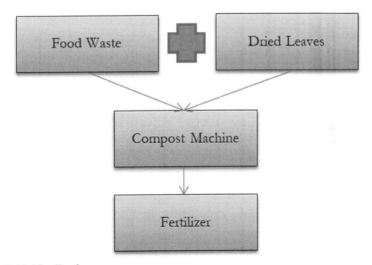

**FIGURE 13.13**   Food compost process.

## 13.3.2   *DATA FOOD WASTE (FW) COMPOSTING*

### 13.3.2.1   *FOOD WASTE (FW) COMPOSTING IN 2018*

During the period from February 2018 to December 2019, the volume of FW collected from the food operators in IIUM Kuantan Campus became a concern to the project. The total production of organic compost in Kuantan Campus for 2018 is 2311 kg. About 60% of the fertilizer was given to IIUM and 40% will be given to DBSB. Figure 13.14 shows the total amount of fertilizer produced by each month in the year of 2018.

Unfortunately, there is no production of fertilizer in January since this composting project start in February. Other than that, in June and November also there is no production of fertilizer because the composting machine was under maintenance. The highest amount production of fertilizer is in July, which is 394.8 kg and the lowest is in February and March, which is 7.5 kg. Since the total amount of wastes processed for composting process is 2311 kg thus the fertilizer produced in the year 2018 is 1413.4 kg. The first half of the year 2018 indicates that it still under does try and error phase. The total reduction achieved is 39%.

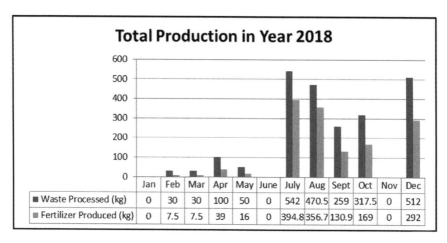

**FIGURE 13.14**   Graph of total waste processed and fertilizer produced in year 2018.

### 13.3.2.1.1   Estimated Fertilizer Revenue for 2018

Since in 2018 the total fertilizer produced 1,413.4 kg, the amount of fertilizer was shared between two organizations. They are DBSB who are actively involved in this project and IIUM Kuantan Campus. About 60% of the total of fertilizer was given to IIUM Kuantan Campus to be used for landscaping in IIUM Kuantan Campus. Meanwhile, another 40% was given to DBSB for marketing. Now, the organic fertilizer is already in the market, and the price is RM 6 per 0.5 kg. Therefore, the estimated fertilizer earnings for the year of 2018 are RM 6784.80.

### 13.3.2.2    FOOD WASTE (FW) COMPOSTING IN 2019

Total waste processed for the first five months in the year 2019 is 21,419 kg and the fertilizer produced is 15,566 kg. Total waste reduction is 27%. Based on the graph shown in Figure 13.15, October recorded the highest value of fertilizer production, which is 1,962 kg because of more waste produce. Meanwhile, January recorded the lowest value of fertilizer production, which is 999 kg. Table 13.2 shows the actual data of wastes collected per month in the year 2019 at IIUM Kuantan Campus.

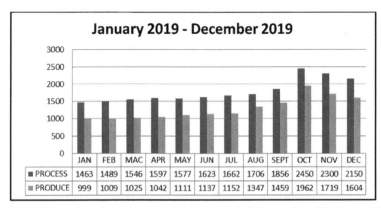

**January 2019 - December 2019**

| | JAN | FEB | MAC | APR | MAY | JUN | JUL | AUG | SEPT | OCT | NOV | DEC |
|---|---|---|---|---|---|---|---|---|---|---|---|---|
| ■ PROCESS | 1463 | 1489 | 1546 | 1597 | 1577 | 1623 | 1662 | 1706 | 1856 | 2450 | 2300 | 2150 |
| ▥ PRODUCE | 999 | 1009 | 1025 | 1042 | 1111 | 1137 | 1152 | 1347 | 1459 | 1962 | 1719 | 1604 |

**FIGURE 13.15**    Graph of total wastes processed and fertilizer produced in the year 2019.

**TABLE 13.2**    Actual Data of Wastes Collected Per Month in Year 2019

| Month | Actual Waste Collected Per Month (kg) | Actual Percentage Per Month (%) | Percentage Increment (%) |
|---|---|---|---|
| January | 1463 | 3.74 | 0.06 |
| February | 1489 | 3.80 | 0.15 |
| March | 1546 | 3.95 | 0.13 |
| April | 1597 | 4.08 | −0.05 |
| Mei | 1577 | 4.03 | 0.12 |
| June | 1623 | 4.15 | 0.1 |
| July | 1662 | 4.25 | 0.11 |
| August | 1706 | 4.36 | 0.38 |
| September | 1856 | 4.74 | 1.52 |
| October | 2450 | 6.26 | −0.39 |
| November | 2300 | 5.87 | −0.38 |
| December | 2150 | 5.49 | |
| | Total Actual Waste Collected = 21419 | Total = 54.72 | Total = 1.75 |

### 13.3.2.2.1    Estimated Fertilizer Revenue for 2019

In 2019 the total fertilizer produced 15,566 kg, the amount of fertilizer was shared between two organizations. They are DBSB who are actively involved in this project and IIUM Kuantan Campus. About 60% of the total of organic compost was given to IIUM Kuantan Campus to be used for landscaping in IIUM Kuantan Campus. Meanwhile, another 40% was given to DBSB for marketing. DBSB is selling the organic fertilizer for RM 6 per 0.5 kg. Therefore, the estimated fertilizer earnings in year 2019 is RM 74,716.80.

As a conclusion, during these 2 years, from 2018 to 2019, IIUM Kuantan Campus has saved cost on chemical fertilizer. Basically, before this composting project was introduced, every 3 months, IIUM Kuantan Campus has used chemical fertilizer for landscaping. The total expenses for the chemical fertilizer are about RM 6500 for every 3 months. Thus, about RM 26,000 was spend only on chemical fertilizer for every year. Then in April 2018, the composting project success and organic fertilizer was produced as the product of a composting process, IIUM Kuantan Campus stopped on acquisition chemical fertilizer. IIUM Kuantan Campus started to use the organic fertilizer for landscaping. As a result, until December 2019, IIUM Kuantan already saves about RM 45,500 on chemical fertilizer.

### 13.3.3    APPLICATION OF ORGANIC FERTILIZER

Currently, landscaping in IIUM Kuantan Campus used organic fertilizer that has been produced by the composting process. Figure 13.16 shows fertilizer's application at IIUM Kuantan Campus.

**FIGURE 13.16**    The application of end product of composting process at IIUM Kuantan Campus.

Since the product produced from a composting process is clean and marketable, the product was given to DBSB for marketing. Figure 13.17 shows the organic fertilizer that ready to use it and sell it.

**FIGURE 13.17**   Organic fertilizer for marketing.

## 13.4   OUTCOME OF APPLICATION OF COMPOSTING PROCESS AT IIUM KUANTAN CAMPUS

The biggest outcome of the composting process at IIUM Kuantan Campus is it helps to lessen the amount of FW and organic waste being disposed to landfills and save precious homeland. Degeneration process of FW in landfills and dumpsites releases methane in the environment that can cause a greenhouse effect. Moreover, the end product of this composting process can be used as an organic fertilizer which is free of chemicals for organic farming. Other than that, one of the outcomes of the application of the composting process at IIUM Kuantan Campus is the potential financial income from the sale of the end product, which is organic fertilizer. The finished product of the composting process in IIUM Kuantan Campus can also be used as compost, which produces high revenue. Figure 13.18 showed the concepts of outcome based on the application of composting process activities.

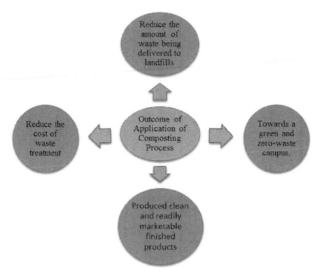

**FIGURE 13.18**  Outcome of application of composting process.

The obtained organic fertilizer product by IIUM has such advantages as no have heavy metals or germs pollution, balanced nutrition (organic matters as well as Nitrogen, Phosphorus, Potassium, and trace metal), high application safety, high efficiency, high bioactivity and long effective time and definitely can improve harvest quality. As landfill space and dumpsite diminish, there will undeniably be more pressure to compost FW along with all organic waste. As tipping fees increase to disposed FW to the landfills, IIUM Kuantan Campus used a composting process, which is a smart financial option as well as a value-added opportunity. Lastly, the benefit of this application of the composting process helps IIUM Kuantan Campus towards a green and zero-waste campus.

## 13.5  CONCLUSION

Based on this, the composting project could be taken as an alternative of waste management operation that is environmentally friendly, economical, sustainable, and wealth-creating. The technique has been used broadly for bioremediation of contaminated soils and sites. However, to maintain composting sustainability, it requires proper handling and suitable technology. Thus, the thesis was designed with the purpose of analyzing the overall management of FWs at IIUM Kuantan Campus. In the beginning, the area affected for this application of the composting process was waste from Mahallah's cafeteria. Currently, the wastes management at Mahallah's cafeterias is well organized. The recycling waste was thrown in the 3R garbage bin. Hence, this application of composting process has been done to organize and improve the current method of waste disposal at Mahallah's cafeteria and other places at IIUM Kuantan Campus, especially for FW where the result from the activity turned into fertilizer and also able to use for all landscapes area in Kuantan campus with very successfully.

## KEYWORDS

- **Food and Agricultural Organization**
- **food and services unit**
- **Islamic International University of Malaysia**
- **key result area**
- **municipal solid waste**
- **sustainable development goals**

## REFERENCES

1. Ishangulyyev, R., Kim, S., & Lee, S. H., (2019). Understanding food loss and waste-why are we losing and wasting food? *Foods (Basel, Switzerland)*, 8(8), 297. Retrieved from: https://doi.org/10.3390/foods8080297
2. Morton, S., Pencheon, D., & Squires, N., (2017). Sustainable development goals (SDGs), and their implementation: A national global framework for health, development and equity needs a systems approach at every level, *British Medical Bulletin, 124*(1), 81–90. Retrieved from: https://doi.org/ 10.1093/bmb/ldx031.
3. Ritchie, H., & Roser, M., (2020). *Urbanization.* Published online at OurWorldInData. org. Retrieved from: https://ourworldindata.org/urbanization (accessed on 22 December 2020).
4. Kathirvale, S., MuhdYunus, M. N., Sopian, K., & Samsuddin, A. H., (2003). Energy potential from municipal solid waste in Malaysia. *Renewable Energy, 29*, 559–567.
5. Tchobanoglous, G., Theisen, H., & Vigil, S. A., (1993). *Integrated Solid Waste Management: Engineering Principle and Management Issue.* McGraw Hill Inc., New York.
6. Yunus, M. N. M., & Kadir, K. A., (2003). The development of solid waste treatment technology based on refuse-derived fuel and biogasification integration. In: *International Symposium on Renewable Energy.* Kuala Lumpur.
7. Wan, A., & Kadir, W. R., (2001). *A Comparative Analysis of Malaysian and the UK.* Addison-Wesley Publishing Company Inc.

8. Hassan, M. N., Awang, M., Afroz, R., & Mohamed, N., (2001). Consumption and impacts on environment: Challenges of globalization. *Paper Presented at the Seminar of Sustainable Consumption: Challenges of Globalization.* Kuala Lumpur, Malaysia.

# CHAPTER 14

# Challenges to Improve Quality of Life with Healthy Food, Less Food Loss, and Waste Reduction

JESÚS ALBERTO GARCÍA GARCÍA[1] and CRISTÓBAL NOÉ AGUILAR[2]

[1]School of Education, Sciences and Humanities, Autonomous University of Coahuila, Saltillo, México, E-mail: jegarciag@uadec.edu.mx

[2]Bioprocesses and Bioproducts Research Group, Food Research Department, School of Chemistry, Autonomous University of Coahuila, Saltillo, México

## ABSTRACT

The aim of this chapter is to present various reflections on the topic of healthy eating and quality of life. In the first moment reflect on the role that the food industry plays in the production of high-calorie food and its degree of social responsibility, in a second moment to analyze the habits and lifestyle of the population in Mexico, in third place reflect on the quality of life model caused by consumerism and finally state the consequences of the disposal of healthy food. Finally, an analysis of the relevance of food loss and waste reduction is made as an integral part of the solutions.

## 14.1 INTRODUCTION

Achieving healthy eating is a multifactorial challenge, because there are economic, cultural, educational, and family components that are decisive in achieving this. For example, many companies in the food industry have a supply-and demand-focused view and base their production on hypercaloric and easy-to-eat foods, such as soft drinks, juices, salted waters, frying, biscuits, sweets, and other treats that are available in many retail establishments.

This suggests that new public policies and greater social engagement from businesses are necessary so that people can access healthier foods and find products available in more retail establishments.

Efforts have been made in Mexico to promote healthy eating of the people; however, they remain insufficient, as a joint effort by social institutions is needed for food-producing companies to produce ultra-processed products that are of high nutritional value that contribute to improving the health of the population. The social commitment of these companies must be greater and put the quality of life of consumers at the center.

Overweight and obesity are health problems that have spread around the world and are not easy to solve, because people are consumerists and find their needs in an industry that supplies products with low nutritional value that who is interested in it is to continue to promote exaggerated consumerism.

Although one-third of the world's food is wasted, human famine continues to increase globally. Under this serious condition, it is essential that every human being becomes a change agent by preventing the waste or breakdown of their food, responsibly deciding the quantities they acquire of food and how they preserve them or, where appropriate, share them.

There is international FAO/UN programs seeking rapprochement with the different governments of countries worldwide in a joint effort to raise awareness of the serious loss and waste of food in some nations while in others, people may starve to death.

In addition to government, such programs can be implemented by industries and productive sectors, civil society, academia, etc., in such a way that policies and regulations can be established to solve the great problem of mishandling of food from its root.

## 14.2   QUALITY OF LIFE WITH HEALTHY FOODS

About the habits and lifestyle of the population in Mexico, they can be inferred from the statistics that exist around overweight and obesity, where it is observed that they have not changed significantly despite the programs and strategies implemented to promote healthier habits and physical activity in the population.

According to the statistical results reported by ENSANUT MC [1], the prevalence of obesity as well as overweight in youth and adult populations has increased over the past five years, particularly in urban areas and suburbs that, in rural areas, where significant increases have also occurred. This behavior occurs widely in different parts of the world. It is noted that this trend is similar in young humans between 12 and 19 years of age, with women being the most affected by the prevalence of overweight and morbid obesity.

Children who are overweight are more likely to have physical health problems such as diabetes and hypertension, whether acute or chronic, as well as consequences on mental health, by promoting the development of anxiety, depression, as well as disorders in eating behavior that may affect your social behavior [2, 6].

According to WHO data [7] statistics on the subject say that since 1975, obesity has nearly tripled a report from the World Health Organization (WHO) in 2016 noted that one-third of overweight people are obese, one-third of who live with morbid obesity, which is a serious health risk factor. Moreover, it is well documented that many people around the world suffer the consequences of overweight and obesity, factors that kill more human lives than those caused by lack of weight and malnutrition. When geo-associating such problems, obesity, and overweight can now be considered a typical problem in countries and regions with high economic incomes.

The habits of many young people are hedonistic, which causes increasingly sedentary behaviors and the adoption of a more consumerist and individualistic lifestyle, where alcohol, cigars, and ultra-processed food are usually consumed. The intake of many of these products is excessive and has harmful effects on the health of children, young people, and adults. The bad eating habits of the population are directly associated with the development of cardio-clarifying and metabolic diseases such as: cholesterol, diabetes, abdominal fat, high blood pressure, and hyperglycemia.

Overweight and obesity are generated by an unhealthy lifestyle and to correct it, the psychological and emotional aspects of a person are an important requirement, since there must be will and discipline to acquire new habits in a highly consumerist society attached to fashions and social styles. Clearly, this can be favored by a responsible industrial food sector and with an educational system that promotes the consumption of food and healthy habits from the early ages of humans.

To consolidate healthy eating, educational strategies need to be adjusted and implemented to educate children to eat nutritious foods, engage in physical activity, and learn about the risks of eating a diet based on consumption of sugars, processed carbohydrates, and sweetened beverages.

An important challenge to overcome is the joint and coordinated work that health professionals, institutions, and governments must develop to improve the food of the population, so it is necessary to form an inter-disciplinary team that allows address ingestions of information about healthy eating, develop campaigns to promote good eating habits and implement programs for the development of eating plans, diets, and weight management.

Another challenge to be addressed is the bad eating habits of the majority of the population, as it is common to see that in the search to meet their food needs many people eat the first thing that is available, they do not have fixed schedules for eating, their snacks are usually soft drinks, frying or biscuits, some do not eat breakfast, or dine very late, among other bad habits that affect your health. This allows us to reflect on the need to educate children, young people and adults in healthy habits based on strategies aimed at developing autonomy, will, discipline, and responsibility in each of the decision-making related to Power.

To adopt a healthy diet, it is important to start with the breakdown of common behaviors such as excessive food intake, unhealthy diets, and transit towards more conscious and responsible consumption, in order to progressively achieve healthier eating. With regard to Quality of Life, it is important to note that as a multidimensional category, it must explain the dimension referred to in its approach; for this reason, this concept analyzes this concept from healthy eating and consequences of the disposal of healthy foods.

## 14.3   FOOD LOSS AND WASTE REDUCTION

It is worth mentioning that there is a great concern since public perception, a great concern about the high rate of waste generation and food waste (FW), so different campaigns of guidance, clarification, information, and dissemination have been implemented in a coordinated manner [3], which are implemented in target 12.3 related to the need to reduce FW in the 2030 agenda for Sustainable Development, at least half of what is currently discarded or lost. It is important to note that the Food and Agriculture Organization estimated in 2011 the need to implement indices that can be used to assess and quantify the reduction of such food losses. These indicators are: food loss rate and FW index. Thus, we can now know that more than 30% of fruit and vegetable products are lost only during cultivation, harvesting, and post-harvesting, and more than 14% of food produced globally is lost before reaching retail markets. The data provided by both indicators are essential to define strategies and policies to prevent FW and loss.

It should be mentioned that food is wasted in different ways, even if it is in excellent consumption conditions but that by the very process of selection and quality, if it does not meet the parameters defined in the markets, the same selection process is the beginning of the waste and loss process. Large amounts of healthy edible foods are often unused or over-used and discarded from domestic cookers and food establishments. Less loss of food and waste would lead to more efficient land use and better management of water resources, which would have a positive effect on livelihoods and in the fight against climate change.

## 14.4   FINAL COMMENTS

It is possible to conclude this analysis by noting that each person must be involved from an early age in the educational and formative processes that lead to a lifestyle that has a favorable impact on reducing food loss and waste. Collaborative programs need to be established between the social, private, academic, and government sectors to raise awareness, develop, and promote policies, strategies, and mechanisms that allow each person to be involved in the prevention of this serious problem.

One of the most important points in the design of strategies for the convergence of food supply chain players, where they must be included from farmers to consumers, to handlers, processors, and traders.

In academia, it is important to educate students on responsible food consumption, changing their attitudes, behaviors, and shopping habits related to individual foods. In addition, the widespread dissemination of schemes for the proper storage of food in households and the understanding of preferential consumption dates to prevent and reduce FW.

If the goal is to create a world of Zero Hunger (SDG 2) by ensuring the modalities of consumption and sustainable food production (SDG 12), it is necessary to re-endo to make the world's population visible, understand, and combat the great problem of food loss and waste generation improving the quality of life with healthy food consumption.

## KEYWORDS

- consumerism
- Food and Agricultural Organization
- harvesting
- sustainable development
- sustainable food production
- World Health Organization

## REFERENCES

1. ENSANUT MC, (2016). *Final Results Report.* Ministry of Health, México.
2. Rodríguez, J., Bastidas, M., Genta, G., & Olaya, P., (2016). Quality of life perceived by overweight and obese schoolchildren from popular sectors of Medellin,, Colombia. *Universitas Psychologica, 15*(2), 301. https://doi.org/10.11144/Javeriana.upsy15-2.cves (accessed on 22 December 2020).
3. AMECA, (2019). *Mexican Association for Food Science.* Red PDA. http://www. amecamex.mx/ (accessed on 22 December 2020).
4. FAO, (2019). *The State of the Food and Agriculture.* E-report. http://www.fao.org/state-of-food-agriculture/en/ (accessed on 22 December 2020).
5. OMS, (2018). *Obesity and Overweight.* Consulted in. https://www.who.int/es/news-room/fact-sheets/detail/obesity-and-overweight (accessed on 22 December 2020).
6. Aranceta, J., Perez-Rodrigo, C., Ribas, L., & Serra-Majem, L. L., (2003). Sociodemographic and lifestyle determinants of food patterns in Spanish children and adolescents: the enKid study. *European Journal of Clinical Nutrition, 57*(1), S40–S44.
7. World Health Organization, (2020). *Obesity and Overweight.* https://www.who.int/news-room/fact-sheets/detail/obesity-and-overweight (accessed on 13 January 2020).

# CHAPTER 15

# UM Zero Waste Campaign (UM ZWC) Waste-to-Wealth Initiative: Promoting Circular Economy at Higher Educational Institutions

SUMIANI YUSOFF,[1] NG CHEE GUAN,[1] MAIRUZ ASMARAFARIZA AZLAN,[2] and NUR SHAKIRAH BINTI KAMARUL ZAMAN[3]

[1]*Institute of Ocean and Earth Science, University Malaya, Kuala Lumpur – 50603, Malaysia, E-mails: sumiani@um.edu.my (S. Yusoff), cheeguan.ng@um.edu.my (Ng. C. Guan)*

[2]*Zero Waste Campaign, University Malaya, Kuala Lumpur – 50603, Malaysia, E-mail: mairuzasmara@gmail.com*

[3]*Institute of Advanced Studies, University Malaya, Kuala Lumpur – 50603, Malaysia, E-mail: shakirahkz90@gmail.com*

## ABSTRACT

With global problems related to waste generation, especially food waste (FW), many higher learning institutions, being one of the primary sources and generators of FW have initiated integrated waste management systems in order to reduce waste generation. With a total of 20 public-funded universities and more than 200 private-funded colleges, it becomes essential for higher educational institutional authorities to reduce the generation of FW. Recycling of FW helps the university to save in disposal cost as well as reduce the environmental burdens. The present chapter aims to showcase the environmental and financial benefits of the University Malaya Zero Waste Campaign (UM ZWC) initiatives as a strategy to promote the circular economy. Avoided carbon emission and monetary saving/gained from UM ZWC initiative from 2012 till 2019 were quantified. Results showed that treating FW and yard waste as a resource could yield large carbon emission reduction. Environmental benefit and financial gained were calculated and compared against business-as-usual. The result of the present study indicates the need to promote the implementation of integrated solid waste management (ISWM) system in higher educational institutions because it not only reduces its institutional carbon footprint and waste disposal costs but also provides the opportunity to more financial gain to promote sustainable circular economy.

## 15.1  INTRODUCTION

### 15.1.1  FOOD WASTE (FW) MANAGEMENT

The escalating trend of the generation of municipal solid waste (MSW) in Malaysia has reached 33,000 tons daily [1]. In general, food waste (FW) contributes to half of the total MSW generated, and approximately more than 70% of it is disposed at landfill sites with higher educational institutions being one of the primary sources of FW in Malaysia [2, 3]. FW is generated throughout the food supply chain, from agricultural production right up to end consumers [4, 5]. At present, landfilling is dominant in Malaysia, accounted for 80% of the MSW collected. Unlike in many European countries, wastes generated from residential, institutional, commercial, and industrial sectors are disposed in landfills directly [6, 7]. At present, there are 186 operating waste disposal sites throughout the nation, out of which only 7 are considered sanitary [8]. Open dumpsites are often in bad conditions and operated without proper protective measures, thus contributing significantly to the climate change effect through methane emission in landfill [6, 9–11]. Moreover, the degradation of FW would affect the communities in the proximity of landfills in respect to health risk and environmental pollution [12]. All these factors have severe impacts on the local environment causing degradation in environmental quality.

### 15.1.2  FOOD WASTE (FW) GENERATED FROM HIGHER EDUCATIONAL INSTITUTIONS

Higher educational institutions with its vast compound which normally house numerous building with different daily activities. Independently of their daily activities, FW is generated from various sources in university campuses, for instance, canteens, hostels, offices, halls, and classrooms. As demand toward adopting sustainability pathways, it has become a necessity to implement waste reduction measures in campus with the aim to reduce the increasing amount of waste generated on campus. Ng and Yusoff [13] reported that a university which houses 17,000 staffs and students produce approximately 4–5 ton of MSW per day, translated to approximately 0.35 kg/capita/day. The measures are in line with waste reduction policies adopted and practiced by other universities globally [14–16]. Adeniran et al. [14] characterized and studied the trend of solid waste generated in the University of Lagos, Nigeria, and found that the recyclable constituting about 75% of the total waste generated, whereas FW makes up 15% of the total waste generated. Gallardo et al. [15] pointed out that waste disposal activities generate environmental impacts. Research work was carried out to the composition, the amount and the distribution of the waste generated in Universitat Jaume, Spain. The study aimed to provide information to the university authorities to propose a set of minimization measures to enhance the current waste management. Previous study by Painter et al. [16] reported a student typically generate 0.5 kg of FW per day. Study found out that FW generation is greatly dependant on numerous factors, for instance, meal times, meal options, distance to the dining hall and the style food is served. Moreover, reducing 10% of FW could help a university to save up to USD 80,000 annually. In Malaysia, daily FW generation by a public-funded university housing 17,000 staffs and students was estimated at about 1–2 ton per day [13]. With a total of 20 public-funded universities and more than 200 private-funded colleges [17], it becomes an essential for higher educational institutional authorities to reduce the generation of FW. Recycling of FW not only helps the universities to save in disposal cost, but also reduce the environmental burdens as well as generating income, thus promoting a circular economy model.

### 15.1.3   *UNIVERSITY MALAYA ZERO WASTE CAMPAIGN (*UM ZWC*)*

In University Malaya (UM), UM Zero Waste Campaign (UM ZWC), a sustainability project has been started in year 2010 aiming for an integrated solid waste management (ISWM) in UM campus. This project runs nine big programs since the inception of the project and continue to strive for more upcoming programs ensuring an integrated waste management system on campus. Figure 15.1 shows the list of the programs by UM ZWC. UM ZWC aims to spearhead the development of an integrated and sustainable waste management model in UM hence reducing the campus environmental burden from its waste generation. The history of UMZWC rooted from a students' class project group, "VeeCYCLE" which developed a recycling project in the Faculty of Engineering with "PRO bin" to promote the best practice of waste segregation at source. The inception of the Green Bag Scheme in 2010 was inspired to promote FW separation and collection in the campus. Subsequently, a composting operation was developed with external funding, support from UM top management as well as technical assistance by Institute for global environmental strategies (IGES), Japan in 2011. Along the years, UM ZWC has established collaboration with various universities, particularly in anaerobic digestion (AD) of FW and compost microbiology research.

Externally, UM ZWC collaborates with lifeline clothing (LLC) Sdn. Bhd. to introduce a used clothes collection and recycling program and TSP Waste Management Sdn. Bhd. for energy recovery of wood waste (WW). UM ZWC also initiated collaboration with the National Solid Waste Management Corporation, Malaysia (SWCorp) on circular economy model. UM ZWC has adopted measures to increase the income for economic sustainability and promote the circular economy concept through the sales of compost and organic produce from its organic farms. From the programs, it is tangibly noticed that there is a reduction in the amount of waste disposed to the landfill and directly reduced campus waste disposal cost. Hence, this present chapter aims to showcase the benefits of the UMZWC initiatives in terms of carbon emission reduction as well as the monetary benefits in promoting circular economy.

### 15.1.4   MULTI-STAGE DIGESTION OF FOOD WASTE (FW)

The combination of AD and composting was carried out as an integrated valorization process of the campus cafeteria FW. Both biogas and high quality compost were produced by such a combined process. Of the various possible treatment technologies for FW, AD certainly plays an important role. Nevertheless, extremely few experiments have been carried out on the AD of the post-consumer FW generated from food outlets. Wang et al. [18] reported that FW shows a high potential to produce methane by AD due to its high organic content. However, many inhibitors, such as accumulation of ammonia and volatile fatty acids (VFAs), usually result in inefficient performances and even process failure. The specific gas production of the methanogenic reactor ranges between 0.75 and 0.88 $m^3$ biogas/kgVS, increases in the removal efficiency of volatile soli [19]. Another study reported the biogas yield for mono-digestion of FW was 0.27–0.642 $m^3$ $CH_4$/kg VS and for the co-digestion of FW with other substrates was 0.272–0.859 $m^3$ $CH_4$/kg VS [20]. Furthermore, experiments have often been made using conventional processes that have now been superseded by knowledge acquired in the field of multiple-stage degradation processes [21–23]. Literature suggested multi-stage anaerobic system provides optimal process stability, increased energy efficacy, and better control over crucial parameters governing assured performance and energy recovery. However, AD produces an effluent that cannot be used directly for land application due to its phytotoxicity [24]. An ultimate bio-oxidative treatment (composting) allows this impediment to be overcome. Study by Coelho et al. [25] showed that anaerobic digestates were found to negatively affect early stages of seed germination, but phytotoxicity effects were decreased by dilution in water. Luo et al. [26] reported that aeration treatment with activated sludge significantly reduced the phytotoxicity of anaerobically digested centrate, as indicated by an enhancement of seed germination index, during the aeration treatment. Alternatively, thermophilic composting of vegetable waste produce compost which satisfied the criteria of maturity level and concentration of hazardous heavy metal was below the threshold value in Korea, thus reduce the phytotoxicity impact to the environment.

Anaerobic pre-treatment was applied prior to the thermophilic composting process in the UM ZWC initiative. FW was digested anaerobically for 30 days, and the digestate was then collected and co-composted with yard waste for 100 days with a mixture ratio of 20:80 (digestate: yard waste). Temperature profile was recorded throughout the composting duration. Although further experimentation is anticipated to study the quality of the compost, it would seem reasonable to suppose that the process of AD may be applied to the FW, where the digestate is later co-composted with yard waste in thermophilic process with satisfactory results. Indeed, the focus point of the co-digestion process is, with a careful choice of mixture condition in thermophilic composting (digestate/yard waste = 20/80), there is the possibility of obtaining performances comparable to those obtained with the single-stage composting of FW. This innovation can represent a significant contribution to the removal of FW as well as generation of multiple by-products (e.g., biogas, and compost). The co-composting of anaerobically digested FW with yard waste is a reliable procedure for an exhaustive management of such putrescible wastes.

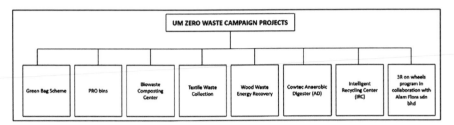

**FIGURE 15.1**  Ongoing projects by UM zero waste campaign since its inception in 2010.

## 15.2   METHODOLOGY

For this study, some of the waste data were collected daily, and some were monthly. The data comprises of FW (kg), green waste (GW) (kg), textile waste (kg), wood waste (WW) (kg), e-waste (EW) (kg), and recyclable material (kg). The data was collected since 2012, and the quantification of carbon reduction from the UM ZWC initiatives from the year 2012 until 2019 was conducted using the carbon footprint assessment.

### 15.2.1   DATA COLLECTION

Data on organic waste, such as FW and garden waste were collected daily. Every day, the FW and garden waste were weighed before they were added to the composting pile. The weight of the organic waste was recorded every day, Monday to Sunday. Meanwhile, data of recyclable materials (plastic, paper, cardboard, aluminum cans) was collected every month, and the collection frequency depends on the amount of the recyclable materials collected. Waste data was collected via several methods, including from the appointed concessionaire and collaborators that will be called upon whenever the recyclable materials at the UM ZWC site reached up to two tons, from the Alam Flora Sdn. Bhd., UM ZWC collaborator that came to UM every month conducting a recycling program, and from each faculty in UM campus. Other waste that was collected upon request is WW and EW. The collaborators of these two types of waste were called when the waste fills up the storage. Meanwhile, the textile waste data was submitted monthly by another UM ZWC collaborator, the LLC.

## 15.2.2  QUANTIFICATION OF GREENHOUSE GASES REDUCTION AND MONETARY BENEFITS

The collected data was analyzed to quantify the potential carbon reduction using carbon footprint assessment. The quantification of the carbon includes the direct and indirect carbon emission from these project initiatives. Scope of direct emission type of carbon covers two aspects, waste avoided from disposed at landfill and total mileage of freight truck avoided, whereas indirect emission category is associated to total energy saved, total diesel saved, total chemical fertilizer consumption avoided and total new products avoided due to recycling effort. The methodology of the calculation for both direct and indirect carbon emissions are summarized in Table 15.1.

**TABLE 15.1**  The Methodology for Carbon Footprint Calculation

| SL. No. | Scope | Formula and Expression |
|---|---|---|
| **Direct Carbon Avoided** | | |
| 1. | Waste avoided from disposed at landfill, $E_1$ | $E_1 = W \times EF_1$ <br> W: waste avoided from transported to landfill (kg) <br> $EF_1$: Disposal of municipal solid waste, sanitary landfill (Default: 0.75451 kg $CO_2$-eq/kg) |
| 2. | Total mileage of freight truck avoided, $E_2$ | $E_2 = (T \times Dt \times 2)\ EF_2$ <br> T: No. of trips reduced <br> Dt: Distance traveled per trip (km) <br> $EF_2$: Transport (Freight lorry 3.5–7.5 metric ton, EURO5) (Default: 0.48337 kg $CO_2$-eq/km) |
| **Indirect Carbon Avoided** | | |
| 1. | Total energy saved, $E_3$ | $E_3 = (E \times EF_{3a}) + [E \times EF_{3b}(GCH_4)] + [E \times EF_{3c}(GN_2O)]$ <br> E: Total energy saved <br> $EF_{3a}$: Emissions per kWh of electricity consumed (Default: 0.770701107 kg $CO_2$-eq/kwh) <br> $EF_{3b}$: Emissions per kWh of electricity consumed (Default: 0.00001131953 kg $CH_4$-eq/kwh) <br> $GCH_4$: Global warming potential for methane gas, $CH_4$ (Default: 23) <br> $EF_{3c}$: Emissions per kWh of electricity consumed (Default: 0.00000695 kg $N_2O$-eq/kWh) <br> $GN_2O$: Global warming potential for Nitrous oxide, $N_2O$ (Default: 300) |

**TABLE 15.1** *(Continued)*

| SL. No. | Scope | Formula and Expression |
|---------|-------|------------------------|
| 2. | Total diesel saved, $E_4$ | $E_4 = (T \times Dt \times Ds \times D \times k) EF_4$ <br> T: <br> No. of trips reduced <br> Ds: <br> Diesel consumption (L) <br> Dt: <br> Distance traveled per trip (km) <br> D: <br> Density of diesel (kg/L) <br> k: no. of trip per day (Default: 2 trips) <br> $EF_4$: Low Sulfur, Diesel Production (Default: 0.45032 kg $CO_2$-eq/kg) |
| 3. | Total chemical fertilizer consumption avoided, $E_5$ | $E_5 = [B \times (1\text{-}M) \times FRF] EF_5$ <br> B: Composted biowaste (kg) <br> M: Mass loss coefficient (Default: 0.85) <br> FRF: Fertilizer replacement factor (Default: 0.5) <br> $EF_5$: Fertilizer production, as N (Default: 2.74053 kg $CO_2$-eq/kg) |
| 4. | Total new products avoided due to recycling effort, $E_6$ | $E_6 = (WT + R + EW) EF_6$ <br> WT: Waste textile (kg) <br> R: Recyclable material (kg) <br> EW: Electronic waste (kg) <br> $EF_6$: Total new products production (average, textiles, papers, plastics, metal, aluminum, etc.) (Default: 1.38488 kg $CO_2$-eq/kg) |

## 15.3 RESULT AND DISCUSSION

### 15.3.1 YEARLY WASTE GENERATION IN UM CAMPUS

Table 15.2 shows the type of waste recycled and treated in the UM campus. This project started with the provision of recycling bins that separate paper, container, and other waste and subsequently expanded to a more diversified type of waste which EW collection bin was provided at UM ZWC site. After this project got funded by CIMB Foundation, there were biological treatments for organic waste collected. FW and GW were biologically treated via composting (CP) with organic compost as the end product. However, not 100% of the collected organic waste is composted; about 7–10% was treated via AD, which produces biogas and liquid compost within 24–48 hours. In 2014, UM ZWC cooperates with LLC Sdn. Bhd. to introduce a used textile (UT) collection and recycling (R) program and TSP Waste Management Sdn. Bhd. for separate collection of WW for energy recovery (ER). In order to expand this project, UM ZWC has collaborated with Alam Flora Sdn. Bhd. on the 3R-On-Wheels program to raise awareness and improve the collection of recyclables (RB).

**TABLE 15.2**   Type of Waste Recycled and Treated by UM ZWC from 2012 Until 2019

| Year | Waste Collected (kg) | | | | | | |
|------|------|------|------|------|------|------|------|
| | **FW** | | **GW** | **UT** | **WW** | **RB** | **EW** |
| | **CP** | **AD** | **CP** | **RC** | **ER** | **RC** | **RC** |
| 2012 | 21,700 | 930 | 930 | – | – | – | – |
| 2013 | 33,010 | 1,530 | 2,960 | – | – | – | 17,740 |
| 2014 | 33,720 | 10,940 | 2,650 | 21,590 | 7,500 | – | 19,320 |
| 2015 | 29,050 | 3,060 | 22,030 | 19,040 | 63,850 | 20,540 | 17,650 |
| 2016 | 55,300 | 3,670 | 5,160 | 19,010 | 45,340 | 51,020 | – |
| 2017 | 69,900 | 3,000 | 21,000 | 15,680 | 52,670 | 49,860 | – |
| 2018 | 69,210 | 700 | 20,340 | 16,920 | 29,520 | 62,700 | – |
| 2019 | 42,210 | 2,690 | 19,560 | 52,270 | – | 81,390 | – |
| Total | 354,100 | 26,520 | 94,630 | 144,510 | 198,880 | 265,510 | 54,710 |

### 15.3.2   QUANTIFICATION OF GREENHOUSE GASES EMISSION AND MONETARY GAIN

Table 15.3 shows the amount of direct and indirect greenhouse gases (GHGs) emission calculated from this study using carbon footprint assessment. Direct carbon emission was calculated from waste avoided from disposed at landfill and total mileage of freight truck avoided. While indirect carbon emission was calculated from total energy saved, total diesel saved, total chemical fertilizer consumption avoided, and total new products avoided due to recycling effort (applicable to papers, plastics, aluminum, metals, and textiles).

**TABLE 15.3**  Cumulative Carbon Emission Avoidance and Monetary Gain from Integrated Solid Waste Management Initiatives by UM ZWC

| SL. No. | Scope | GHG Emission Avoided (kg CO$_2$eq) | Cost Avoided/ Gained (RM) |
|---------|-------|-----------------------------------|---------------------------|
| 1. | Waste avoided from disposed at landfill | 826,837 | 284,670 |
| 2. | Total mileage of freight truck avoided[1] | 41,493 | – |
| 3. | Total energy saved | 3,323,788 | 897,422 |
| 4. | Total diesel saved | 2,573 | 10,344 |
| 5. | Total chemical fertilizer consumption avoided | 84,510 | 168,275 |
| 6. | Total new products avoided due to recycling effort[2] | 638,291 | – |
| 7. | Sales of compost | – | 42,000 |
| 8. | Sales of recyclables (papers, plastics, aluminum, metals, textiles) | – | 18,585 |
| 9. | Sales of organic produce | – | 500 |
|  | Total | 4,917,492 | 1,421,796 |

The avoided carbon emission from UM ZWC initiative from 2012 till 2019 was quantified. Carbon emissions avoided from landfill is associated to the total amount of mixed recyclable materials collected and diverted away from landfill. Other carbon emissions avoided were associated to the total amount of mileage to landfill avoided, total amount of diesel avoided due to hauling trip reduction, energy consumption reduction due to the utilization of biogas produced, total amount of N-fertilizer replaced by compost from UM ZWC and total amount of new product replaced by recyclable materials collected from UM ZWC recycling program. The percentage of GHG emission reduction by source is illustrated in Figure 15.2. The largest carbon emission is avoided through utilization of biogas to replace energy which accounts for 67%, followed by avoided emission from FW disposal at landfill (17%) and recycling effort (13%). Carbon emissions reduction due to mileage of truck avoided, diesel consumption avoided and substitution of chemical fertilizer are minimal.

Revenue is generated by selling compost at RM 5/kg. By using open-air Takakura method, from record, 1 kg of FW produce 0.15 kg of compost after 60 days. The weight reduction is due to loss of moisture content. The disposal cost includes the hauling charge of RM 250/haul and landfill tipping fee of RM 55/ton. Figure 15.3 shows that the biggest saving in

associated with the saving of energy consumption, accounted for 63% of the total monetary gain, followed by disposal cost saving of FW (20%) and saving in chemical fertilizer (12%). Total saving in diesel consumption as well as revenue gained from sales of compost and RB are relatively insignificant.

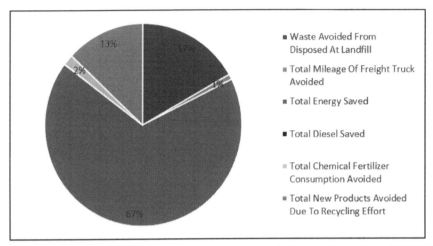

**FIGURE 15.2**   The percentage of cumulative carbon emission avoidance from UMZWC initiatives from 2012 till 2019.

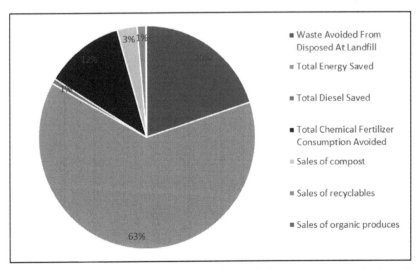

**FIGURE 15.3**   The percentage breakdown of cumulative monetary saved/gained from UMZWC initiatives from 2012 till 2019.

### 15.3.3   COMPOSTING AS AN ALTERNATIVE TREATMENT OPTION

Composting is a cheaper and simpler technology in solid waste management (SWM), and has played an important role in treating, minimizing, and utilizing organic wastes produced by municipalities, agricultural, and agro-industrial activities [27]. Compost has been proven to contribute lower emissions of greenhouse gases to the environment as compared to chemical fertilizers [28]. However, some arguments have arisen over the years questioning its impact on the environment through gaseous emissions and impurities released from the process. Even though composting is known for its lower environmental impacts compared to landfill and incineration, carbon emissions still occur during transporting and processing [29]. Before composting starts, wastes are transported to composting sites in trucks, which are major contributors of emission and production of greenhouse gases, mainly carbon dioxide. It is common knowledge that carbon dioxide is one of the main causes of global warming and climate change [30]. Hence, campus-based composting is feasible in order to reduce the transportation of the waste materials. Moreover, campus-based composting plant is smaller in scale where passive-aerated static pile method is a popular practice, hence the environmental impacts may vary [29, 31]. Industrial composting, in-vessel composting or aerated windrow composting, imply the consumption of energy for waste handling and processing, the emission of odors and other contaminants, and the mixture of different quality materials. On the contrary, campus-based composting presents some potential benefits when compared to the industrial process. For instance, decentralized community composting avoid high usage of energy and reduce the methane generation from the smaller composting pile.

The composting center in UM produces compost from FW and yard waste generated from residential colleges and restaurants in the campus. The operation of the composting center is shown in Figure 15.4. The process begins with the collection and delivery of feedstock on a daily basis (approximately 300 kg of food waste/day) from the restaurants and residential colleges to the composting site (5.2 km). Inorganic fractions are removed manually. Yard waste is shredded into smaller pieces using a 6-HP wood chipper with a capacity of 100 kg/h, subsequently layered on the ground as the base with a thickness of 15 cm to cover an area of about 3 m in diameter. FW digestate is then poured on to the as-prepared base and covered with the YW at a FW-to-YW weight ratio of 2:8. This process is repeated until the pile reaches the height of 1.5 m. The piles are turned manually every week to allow aeration for 100 days. Then, the stabilized compost is then transferred to an open area for drying in another two weeks. After the compost is dried, it is ground with a 12-HP grinder (SIMA/FG-400 x 200) and packed into sacks manually, sewn, and weighted before being utilized in the campus landscape to replace the use of chemical fertilizer.

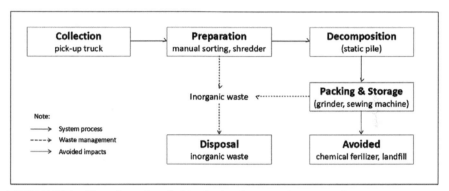

**FIGURE 15.4**    Flow chart of studied institutional community composting systems.

The main benefits of UM ZWC initiatives stemmed from avoided emissions from the avoided activity to dispose the FW and the substitute of chemical fertilizer and energy consumption. Results showed that treating FW and yard waste as a resource could yield large carbon emission reduction. Environmental impacts were calculated and compared between Business-as-Usual and UM ZWC initiatives. Subsequently, composting of FW also promotes soil carbon sequestration. For instance, the application of compost to land has helped to store a large amount of carbon below ground, where it becomes soil organic matter to help promote plant growth, subsequently increasing more carbon uptake from the atmosphere [32, 33].

## 15.4   CONCLUSION

Since the inception of UM ZWC integrated waste management initiatives, a total of saving of 4,917,492 kg $CO_2$eq emission and RM 1,421,796 of monetary benefit have been achieved. The saving can be translated to approximately 8.4 kg $CO_2$eq per student per year and RM 10 per student per year, respectively. AD of FW prior to thermophilic composting process is a promising waste management alternative due to multiple by-products generated. This study evaluated the environmental and economic benefits of multi-stage digestion method of FW and yard waste using the UM ZWC initiative as a case study. Results were compared with Business-as-Usual of conventional waste management practices (landfill). The results showed that an institutional sustainability initiative such as UM ZWC exhibits great GHG emission reduction, thus significantly reduce the environmental impacts associated with global warming potential. In terms of economics, UM ZWC had shown the monetary gain in respect to disposal cost avoidance, land acquisition, substitute of virgin materials from recycling, and avoided environmental burden. The results exhibited from UM ZWC initiatives show a promising feasible and practical circular economy model for managing increasing institutional FW generation in terms of both environmental and economic sustainability.

## ACKNOWLEDGMENTS

This research is made possible by the supported grants under Eco-Campus University Malaya, Kuala Lumpur (ZWC Living Lab Grant) [LL004-15SUS] and Research University Grant-UMCares [MRUN2019-2C].

## KEYWORDS

- anaerobic digestion
- carbon avoidance
- circular economy
- e-waste
- food waste
- global environmental strategies
- higher educational institutions
- multi-stage codigestion

## REFERENCES

1. Moh, Y. C., & Manaf, L. A., (2014). Overview of household solid waste recycling policy status and challenges in Malaysia. *Resources, Conservation and Recycling, 82*, 50–61.
2. Bong, C. P. C., Ho, W. S., Hashim, H., Lim, J. S., Ho, C. S., Tan, W. S. P., & Lee, C. T., (2017). Review on the renewable energy and solid waste management policies

towards biogas development in Malaysia. *Renewable and Sustainable Energy Reviews, 70*, 988–998.

3. Moh, Y., (2017). Solid waste management transformation and future challenges of source separation and recycling practice in Malaysia. *Resources, Conservation and Recycling, 116*, 1–14.

4. Thyberg, K. L., & Tonjes, D. J., (2016). Drivers of food waste and their implications for sustainable policy development. *Resources, Conservation and Recycling, 106*, 110–123.

5. Gustavsson, J., Cederberg, C., Sonesson, U., Van, O. R., & Meybeck, A., (2011). Global Food Losses and Food Waste-FAO Report. *Food and Agriculture Organization (FAO) of the United Nations*, 1–37.

6. Ismail, S. N. S., & Latifah, A. M., (2013). The challenge of future landfill: A case study of Malaysia. *Journal Toxicology and Environmental Health Sciences (JTEHS), 5*(3), 2400–2407.

7. Tang, K. H. D., (2019). Climate change in Malaysia: Trends, contributors, impacts, mitigation and adaptations. *Science of the Total Environment, 650*, 1858–1871.

8. Hoo, P. Y., Hashim, H., & Ho, W. S., (2018). Opportunities and challenges: Landfill gas to biomethane injection into natural gas distribution grid through pipeline. *Journal of Cleaner Production, 175*, 409–419.

9. Venkat, K., (2011). The climate change and economic impacts of food waste in the United States. *International Journal on Food System Dynamics, 2*(4), 431–446.

10. De, S., Maiti, S., Hazra, T., Debsarkar, A., & Dutta, A., (2016). Leachate characterization and identification of dominant pollutants using leachate pollution index for an uncontrolled landfill site. *Global Journal of Environmental Science and Management, 2*(2), 177–186.

11. Lou, X. F., & Nair, J., (2009). The impact of landfilling and composting on greenhouse gas emissions: A review. *Bioresource Technology, 100*(16), 3792–3798.

12. Palmiotto, M., Fattore, E., Paiano, V., Celeste, G., Colombo, A., & Davoli, E., (2014). Influence of a municipal solid waste landfill in the surrounding environment: Toxicological risk and odor nuisance effects. *Environment International, 68*, 16–24.

13. Ng, C. G., & Yusoff, S., (2015). Assessment of GHG emission reduction potential from source-separated organic waste (SOW) management: Case study in a higher educational institution in Malaysia. *Sains Malays, 44*(2), 193–201.

14. Adeniran, A. E., Nubi, A. T., & Adelopo, A. O., (2017). Solid waste generation and characterization in the University of Lagos for a sustainable waste management. *Waste Management, 67*, 3–10.

15. Gallardo, A., Edo-Alcón, N., Carlos, M., & Renau, M., (2016). The determination of waste generation and composition as an essential tool to improve the waste management plan of a university. *Waste Management, 53*, 3–11.

16. Painter, K., Thondhlana, G., & Kua, H. W., (2016). Food waste generation and potential interventions at Rhodes University, South Africa. *Waste Management, 56*, 491–497.

17. Grapragasem, S., Krishnan, A., & Mansor, A. N., (2014). Current trends in Malaysian higher education and the effect on education policy and practice: An overview. *International Journal of Higher Education, 3*(1), 85–93.

18. Wang, P., Wang, H., Qiu, Y., Ren, L., & Jiang, B., (2018). Microbial characteristics in anaerobic digestion process of food waste for methane production: A review. *Bioresource Technology, 248*, 29–36.

19. Micolucci, F., Gottardo, M., Pavan, P., Cavinato, C., & Bolzonella, D., (2018). Pilot-scale comparison of single and double-stage thermophilic anaerobic digestion of food waste. *Journal of Cleaner Production, 171,* 1376–1385.
20. Bong, C. P. C., Lim, L. Y., Lee, C. T., Klemeš, J. J., Ho, C. S., & Ho, W. S., (2018). The characterization and treatment of food waste for improvement of biogas production during anaerobic digestion: A review. *Journal of Cleaner Production, 172,* 1545–1558.
21. Srisowmeya, G., Chakravarthy, M., & Devi, G. N., (2019). Critical considerations in two-stage anaerobic digestion of food waste: A review. *Renewable and Sustainable Energy Reviews,* 109587.
22. Li, W., Loh, K. C., Zhang, J., Tong, Y. W., & Dai, Y., (2018). Two-stage anaerobic digestion of food waste and horticultural waste in high-solid system. *Applied Energy, 209,* 400–408.
23. De Gioannis, G., Muntoni, A., Polettini, A., Pomi, R., & Spiga, D., (2017). Energy recovery from one-and two-stage anaerobic digestion of food waste. *Waste Management, 68,* 595–602.
24. Enaime, G., Baçaoui, A., Yaacoubi, A., Belaqziz, M., Wichern, M., & Lübken, M., (2020). Phytotoxicity assessment of olive mill wastewater treated by different technologies: Effect on seed germination of maize and tomato. *Environmental Science and Pollution Research,* 1–12.
25. Coelho, J. J., Prieto, M. L., Dowling, S., Hennessy, A., Casey, I., Woodcock, T., & Kennedy, N., (2018). Physical-chemical traits, phytotoxicity and pathogen detection in liquid anaerobic digestates. *Waste Management, 78,* 8–15.
26. Luo, W., Zhang, B., Bi, Y., Li, G., & Sun, Q., (2018). Effects of sludge enhanced aeration on nutrient contents and phytotoxicity of anaerobically digested centrate. *Chemosphere, 203,* 490–496.
27. Cerda, A., Artola, A., Font, X., Barrena, R., Gea, T., & Sánchez, A., (2018). Composting of food wastes: Status and challenges. *Bioresource Technology, 248,* 57–67.
28. Andersen, J. K., Boldrin, A., Christensen, T. H., & Scheutz, C., (2012). Home composting as an alternative treatment option for organic household waste in Denmark: An environmental assessment using life cycle assessment-modeling. *Waste Management, 32*(1), 31–40.
29. Martínez-Blanco, J., Lazcano, C., Christensen, T. H., Muñoz, P., Rieradevall, J., Møller, J., & Boldrin, A., (2013). Compost benefits for agriculture evaluated by life cycle assessment: A review. *Agronomy for Sustainable Development, 33*(4), 721–732.
30. Clark, S., Khoshnevisan, B., & Sefeedpari, P., (2016). Energy efficiency and greenhouse gas emissions during transition to organic and reduced-input practices: Student farm case study. *Ecological Engineering, 88,* 186–194.
31. Rasapoor, M., Adl, M., & Pourazizi, B., (2016). Comparative evaluation of aeration methods for municipal solid waste composting from the perspective of resource management: A practical case study in Tehran, Iran. *Journal of Environmental Management, 184,* 528–534.
32. Ryals, R., Hartman, M. D., Parton, W. J., DeLonge, M. S., & Silver, W. L., (2015). Long-term climate change mitigation potential with organic matter management on grasslands. *Ecological Applications, 25*(2), 531–545.
33. Boldrin, A., Andersen, J. K., Møller, J., Christensen, T. H., & Favoino, E., (2009). Composting and compost utilization: Accounting of greenhouse gases and global warming contributions. *Waste Management and Research, 27*(8), 800–812.

34. Kim, E. Y., Hong, Y. K., Lee, C. H., Oh, T. K., & Kim, S. C., (2018). Effect of organic compost manufactured with vegetable waste on nutrient supply and phytotoxicity. *Applied Biological Chemistry*, *61*(5), 509–521.

## (FOOTNOTES)

1  The cost-saving from total mileage is included into the disposal cost avoided in association to the waste avoided from landfill.
2  The cost-saving from recycling is excluded due to the limitation of data.

# CHAPTER 16

# Bio-Compost Production from Organic Waste: A Brief Review

M. R. M. HUZAIFAH,[1] S. ISMAIL,[2] N. A. UMOR,[3] Z. JOHAR,[4] S. Y. ANG,[4] and I. M. MEHEDI[5]

[1]Department of Crop Science, Faculty of Agriculture, Universiti Putra Malaysia (UPM), 43400 – UPM Serdang, Selangor, Malaysia, E-mail: mdhuzaifahroslim@gmail.com

[2]School of Ocean Engineering, Universiti Malaysia Terengganu, 21030 – Kuala Nerus, Terengganu, Malaysia

[3]School of Biology, Universiti Teknologi MARA (UiTM) Negeri Sembilan, Kampus Kuala Pilah, Pekan Parit Tinggi, 72000 – Kuala Pilah, Negeri Sembilan, Malaysia

[4]National Hydraulic Research Institute of Malaysia NAHRIM, 43300 – Seri Kembangan, Selangor, Malaysia

[5]Center of Excellence in Intelligent Engineering Systems (CEIES), King Abdulaziz University, Jeddah, Saudi Arabia

## ABSTRACT

Organic waste is one of the environmental concerns nowadays. The organic waste production in Malaysia reaches thousands of tons a day, and this leads to organic waste management should be taken seriously. Composting seems to be the best solution to solve this problem. This article is a review on the production of household waste in Malaysia and the related study on food waste (FW) composting.

## 16.1 INTRODUCTION

To date, organic waste management is becoming one of the major issues globally. Unsystematic management on a large amount of organic waste will have an adverse effect on the environment. According to Abd. Hamid et al. [1], out of the total waste generated at waste disposal sites in Malaysia, 48–68% is consisting of organic waste. Normal practice and a more suitable solution in Malaysia to manage organic waste by using landfill, incineration, and decaying organic wastes by using biological processes [2, 3]. The advantage of using biological process as one of the decomposition methods are low cost and can generate income. This is because the composting process is carried out by microbial activity. Several parameters influence the effectiveness of this method, such as pH, temperature, moisture content, aeration, and C:N ratio.

Shymala and Belagali [5] stated that one of the recommended alternatives for solid waste management (SWM) is to recycle or compost organic waste into valuable products such as fertilizer to prevent the increasing of waste. The final product of composting can be useful for soil fertility, for instance soil conditioner and organic fertilizer due to higher contain of nutrient for the plant [6]. In addition, public awareness of the disadvantages of using chemical poisons has made organic food in high demand. The presence of microbial community, such as bacteria, fungi, and worms in compost help to stabilize organic matters.

Besides that, to obtain the best results in composting production, it depends on the substrate used and the physical conditions such as humidity, temperature, and ventilation of where the composting process takes place. According to Ahmad et al. [7], suitable waste, which is biodegradable material, must be used in order to facilitate the composting process. There are some benefit from applying the composting process in SWM, such as reduce volume, weight, and water content of the waste generated [8, 9]. In the agriculture sector, compost production can be used to improve the soil texture, soil condition, provided nutrient to the plant, and reduce dependence on chemical fertilizers water infiltration rate, water holding capacity, and tilth [10, 11].

## 16.2   OVERVIEW OF ORGANIC WASTE IN COMPOSTING

Across the world, most of the organic waste will end up in landfills. In many disposal methods, composting might be the best alternative to overcome this problem. A. A. Kadir et al. [12] has mentioned that organic waste can be separated into three classifications, which are agriculture waste, municipal solid waste (MSW), and kitchen waste. This chapter will focus on how the kitchen waste were disposed of using the composting method.

### 16.2.1   HOUSEHOLD WASTE

The National Solid Waste Management Department (NSWMD) carried a study on solid waste composition, characteristic, and existing practice of solid waste recycling in Malaysia on year 2012 [13]. From the survey, total household waste generation for peninsular Malaysia was approximately 18,000 metric tons per day as shown in Tables 16.1 and 16.2. From that value, the value of waste generated is about 0.8 kg/capita/day. The average waste generated in urban areas (0.83 kg/capita/day) is higher than waste generated in rural areas (0.73 kg/capita/day). From the results, the trend of waste generated is influenced by housing types which are housing types medium and high cost is higher than the low-cost housing types.

Table 16.3 indicated the household waste generation in Malaysia. The addition of the population from Sabah and Sarawak approximately 6 million people reduced the marginal rate household waste generation in Malaysia. It is caused by the generation of household waste in Sabah and Sarawak is lower than the generation of waste from Peninsular Malaysia. The per capita household waste generation rate for Malaysia is 0.76 kg/capita/day, which is slightly lower than that of the rate in Peninsular Malaysia (0.8 kg/capita/day).

In terms of strata, rural household generated low waste (0.68 kg/capita/day) compared to urban area (0.8 kg/capita/day). In terms of housing type, the pattern follows that of Peninsular Malaysia, where the per capita household waste generation rate for medium-high cost housing types is higher than the low-cost housing types.

### 16.2.2   FOOD WASTES (FWS)

Based on the research done by Saeed et al. [14], the food waste (FW) contributes 60% from total MSW generated at the landfills. Another problem will arise if FW is not managed properly, which can cause odor pollution. A. A. Kadir et al. [15] handled a research on FW composting from Makanan Ringan Mas using backyard composting with 6 reactors (Table 16.4). Fermentation liquid and 3 kg fermented bed were mixed and were put into each reactor. The aim of this research is to identify the physical and chemical parameters of composting FW from Makanan Ringan Mas.

TABLE 16.1 Average Household Waste Generation in 2012 for Peninsular Malaysia [13]

| | Peninsular Malaysia | | | | | | | | |
| Housing Type | Urban | | | Rural | | | Overall | | |
| | Population | Per Capita (kg/capita/day) | Total (MT/day) | Population | Per Capita (kg/capita/day) | Total (MT/day) | Population | Per Capita (kg/capita/day) | Total (MT/day) |
|---|---|---|---|---|---|---|---|---|---|
| Low cost Landed | 2,284,650 | 0.78 | 1,772 | 1,395,530 | 0.73 | 1,024 | 3,680,180 | 0.76 | 2,797 |
| Low cost High-rise | 3,279,077 | 0.65 | 2,139 | 452,967 | 0.77 | 350 | 3,732,044 | 0.67 | 2,490 |
| Medium cost Landed | 6,888,828 | 0.93 | 6,414 | 2,298,782 | 0.72 | 1,647 | 9,187,610 | 0.88 | 8,061 |
| High-Medium cost High-rise | 2,012,187 | 0.91 | 1,826 | – | – | – | 2,012,187 | 0.91 | 1,826 |
| High cost Landed | 2,526,676 | 0.76 | 1,933 | 1,430,647 | 0.72 | 1,023 | 3,957,324 | 0.75 | 2,956 |
| Total | 16,991,419 | 0.83 | 14,083 | 5,577,926 | 0.73 | 4,045 | 22,569,345 | 0.80 | 18,129 |

TABLE 16.2 Average Household Waste Generation in 2012, Sabah, and Sarawak [13]

| Housing Type | Sabah and Sarawak | | | | | | | | |
|---|---|---|---|---|---|---|---|---|---|
| | Urban | | | Rural | | | Overall | | |
| | Population | Per Capita (kg/capita/day) | Total (MT/day) | Population | Per Capita (kg/capita/day) | Total (MT/day) | Population | Per Capita (kg/capita/day) | Total (MT/day) |
| Low cost Landed | 388,369 | 0.59 | 229 | 618,650 | 0.61 | 375 | 1,007,019 | 0.60 | 604 |
| Low cost High-rise | 488,638 | 0.49 | 241 | 403,683 | 0.61 | 244 | 892,321 | 0.54 | 486 |
| Medium cost Landed | 1,279,249 | 0.62 | 796 | 1,077,513 | 0.58 | 629 | 2,356,762 | 0.60 | 1,425 |
| High-Medium cost High-rise | 352,379 | 0.73 | 256 | — | — | 352,379 | 0.73 | 256 | — |
| High cost Landed | 624,916 | 0.61 | 380 | 531,393 | 0.61 | 326 | 1,156,309 | 0.61 | 706 |
| Total | 3,133,551 | 0.61 | 1,902 | 2,631,239 | 0.60 | 1,575 | 5,764,790 | 0.60 | 3,477 |

**TABLE 16.3**  Average Household Waste Generation in 2012, Malaysia [13]

| Housing Type | Malaysia | | | | | | | | | | | |
| | Urban | | | Rural | | | Overall | | | | | |
| | Population | Per Capita (kg/capita/day) | Total (MT/day) | Population | Per Capita (kg/capita/day) | Total (MT/day) | Population | Per Capita (kg/capita/day) | Total (MT/day) |
| --- | --- | --- | --- | --- | --- | --- | --- | --- | --- |
| Low cost Landed | 2,675,954 | 0.74 | 1,988 | 2,019,579 | 0.69 | 1,397 | 4,695,533 | 0.72 | 3,384 |
| Low cost High-rise | 3,778,052 | 0.63 | 2,394 | 830,781 | 0.71 | 586 | 4,608,833 | 0.65 | 2,981 |
| Medium cost Landed | 8,167,292 | 0.89 | 7,245 | 3,377,231 | 0.67 | 2,276 | 11,544,523 | 0.82 | 9,521 |
| High-Medium cost High-rise | 2,366,232 | 0.89 | 2,095 | – | – | 2,366,232 | 0.89 | 2,095 | 2,366,232 |
| High cost Landed | 3,137,440 | 0.73 | 2,303 | 1,981,574 | 0.68 | 1,343 | 5,119,014 | 0.71 | 3,137,440 |
| **Total** | **20,124,970** | **0.80** | **16,025** | **8,209,165** | 0.68 | 5,601 | 28,334,135 | 0.76 | 20,124,970 |

**TABLE 16.4**   Classification and Types of Food Waste [15]

| Reactor | Classification of Food Waste | Types of Food Waste |
|---|---|---|
| Control | Control | – |
| A | Processed food waste | Chips + candy |
| B | Raw food waste | Banana peel + tapioca peel + grated coconut |
| A+B | Processed food waste + raw food waste | Candy + chips + tapioca peel + banana peel + grated coconut |
| C | Processed food waste + raw food waste (Product A) | Candy + grated coconut |
| D | Processed food waste + raw food waste (Product B) | Chips + tapioca peel + banana peel |

The highest temperature recorded was 52°C, and this showed that the microorganisms such as bacteria and fungi are responsible for heat generation during composting cause in lost moisture content [16]. Comparison between chemical fertilizer and FW compost in term of NPK composition were shown in Table 16.5. Based on the table, there are huge gap in NPK between chemical fertilizer and compost from FW. Other studies were done by Saad et al. [17] on composting of mixed yard and FWs with effective microbes. From the results, it can be concluded that the nutrient that being produced varied according to the type or organic compost and wastes that was used (Table 16.6). Even though chemical fertilizers have a high percentage in NPK, organic fertilizers are still the best fertilizers because of their micronutrient content and their ability to be both fertilizer and regenerate poor soil condition.

**TABLE 16.5**   Comparison of Nitrogen, Phosphorus, and Potassium with Chemical Fertilizer [15, 17]

| Sample | Nitrogen (N) (%) | Phosphorus (P) (%) | Potassium (K) (%) |
|---|---|---|---|
| Chemical fertilizer | 20 | 40 | 25 |
| Reactor control | 2.93 | 0.020 | 0.041 |
| Reactor A | 3.92 | 0.023 | 0.045 |
| Reactor B | 4.85 | 0.041 | 0.155 |
| Reactor A+B | 4.68 | 0.027 | 0.057 |
| Reactor C | 4.39 | 0.025 | 0.050 |
| Reactor D | 4.00 | 0.027 | 0.070 |

**TABLE 16.6**   Chemical Analysis for Organic Compost [18]

| Sample | Nitrogen (N) (%) | Phosphorus (P) (%) | Potassium (K) (%) |
|---|---|---|---|
| Cow dung | 1.30 | 0.58 | 2.15 |
| Poultry manure | 2.21 | 2.98 | 2.05 |
| Cassava peelings compost | 1.70 | 0.86 | 1.50 |
| Rabbit droppings | 1.04 | 0.99 | 2.05 |
| Cane rat droppings | 1.95 | 2.06 | 3.30 |

Pathak et al. [19] conducted a research on composting kitchen waste to determine physicochemical and microbiological properties using bioreactor. The samples were analyzed every 15 days until 135 days for four consecutive years. At the early stage, the pH was low due to the fermentation of carbohydrate that release organic acids. Then, after 15 days, the pH was increase when ammonification took place, and the pH remains constant at 6.75 as a result of accumulation of organic acid during degradation. The highest temperature recorded was 64°C between days 35 to 45 and also affected the moisture content in biocomposter when the moisture content drops to 21.7% from 55.8%. This condition was influenced by the microbial activities that show an increase of microbial count at an early stage and declined with composting age. The nutrients content in this kitchen waste compost were in the range of 1.16 to 1.20% for nitrogen, 0.030 to 0.053% for phosphorus and 0.297 to 0.377% for potassium.

In 2011, another research on microbial composting was conducted by Arslan [8] to find out the effect of aeration rate on composting of vegetable and fruit wastes. Different aeration rate was used which are 0.37, 0.49, 0.62, 0.74, 0.86 and 0.99 L/min kg VS to compost the waste for 18 days. The reactor with the highest aeration shows the lowest moisture content. It is believed that the excess aeration caused the drying of the compost. For temperature, the lowest aeration reactor (reactor 1) resulted in the highest temperature compared to other reactors. In this study, the author suggested that the optimum aeration rate for aerobic composting is 0.62 L/min kg VS.

## 16.3   CONCLUSION

Based on a previous study, it shows that the composting process is one of the best methods to reduce FW generation at the landfills since composting can give necessary nutrients needed for agriculture and can be used as organic fertilizer as a substitute to chemical fertilizer. The comparison with common chemical fertilizer shows that FW composting from Makanan Ringan Mas determines lower NPK value. However, the comparison between FWs composting with available organic compost only shows a slight difference in NPK value. Nevertheless, organic composting is recommended because this method causes less harm to the environment in order to solve the waste issue.

## KEYWORDS

- **composting process**
- **fertilizer**
- **food waste**
- **National Solid Waste Management Department**
- **organic compost**
- **solid waste management**

## REFERENCES

1. Hamid, A. A., Ahmad, A., Ibrahim, M. H., & Rahman, N. N. N. A., (2012). Food waste management in Malaysia-current situation and future management options. *Journal of Industrial Research and Technology*, 2(1), 36–39.
2. Dhokhikah, Y., & Trihadiningrum, Y., (2012). Solid waste management in Asian developing countries: Challenges and opportunities. *Journal of Applied Environmental and Biological Sciences*, 2(7), 329–335.
3. Rawat, M., Ramanathan, A. L., & Kuriakose, T., (2013). Characterization of Municipal Solid Waste Compost (MSWC) from Selected Indian Cities: A Case Study for its Sustainable Utilization. *Journal of Environmental Protection*, 4(2), 163–171.
4. Fathi, H., Zangane, A., Fathi, H., Moradi, H., & Lahiji, A. A., (2014). Municipal solid waste characterization and it is assessment for potential compost production: A case study in Zanjan city, Iran. *American Journal of Agriculture and Forestry*, 2(2), 39–44.
5. Shyamala, D. C., & Belagali, S. L., (2012). Studies on variations in physicochemical and biological characteristics at different maturity stages of municipal solid waste compost. *International Journal of Environmental Sciences*, 2(4), 1984–1997.
6. Rama, L., & Vasanthy, M., (2014). Market waste management using compost technology. *International Journal of Plant, Animal and Environmental Sciences, (IJPAES)*, 4(4), 57–61.
7. Ahmad, R., Jilani, G., Arshad, M., Zahir, Z. A., & Khalid, A., (2007). Bio-conversion of organic wastes for their recycling in agriculture: An overview of perspectives and prospects. *Annals of Microbiology*, 57(4), 471–479.

8. Arslan, E. I., Ünlü, A., & Topal, M., (2011). Determination of the effect of aeration rate on composting of vegetable-fruit wastes. *Clean-Soil, Air, Water, 39*(11), 1014–1021.

9. Hargreaves, J. C., Adl, M. S., & Warman, P. R., (2008). A review of the use of composted municipal solid waste in agriculture. *Agriculture, Ecosystems and Environment, 123*(1–3), 1–14.

10. Hernández, A., Castillo, H., Ojeda, D., Arras, A., López, J., & Sánchez, E., (2010). Effect of vermicompost and compost on lettuce production. *Chilean Journal of Agricultural Research, 70*(4), 583–589.

11. Pangnakorn, U., (2006). Valuable added the agricultural waste for farmers using in organic farming groups in Phitsanulok, Thailand. *Proceeding of the Prosperity and Poverty in a Globalized World-Challenges for Agricultural Research*. Bonn, Germany.

12. Risse, M., & Faucette, B., (2009). *Food Waste Composting: Institutional and Industrial Applications* (pp. 1–8). The University of Georgia, The Cooperative Extension Service Bulletin – 1189.

13. Jabatan Pengurusan Sisa Pepejal Negara, (2014). *Survey on Solid Waste Composition, Characteristics and Existing Practice of Solid Waste Recycling in Malaysia*. In: Kementerian Kesejahteraan Bandar, Perumahan Dan Kerajaan Tempatan.

14. Saeed, M. O., Hassan, M. N., & Mujeebu, M. A., (2008). Development of municipal solid waste generation and recyclable components rate of Kuala Lumpur: Perspective study. In: *International Conference on Environment (ICENV 2008)* (pp. 15–17). Penang, Malaysia.

15. Kadir, A. A., Ismail, S. N. M., & Jamaludin, S. N., (2016). Food waste composting study from *Makanan ringan* Mas. In: *IOP Conference Series: Materials Science and Engineering* (Vol. 136, No. 1, p. 012057). IOP Publishing.

16. Tweib, S. A., Rahman, R. A., & Khalil, M. S., (2011). Composting of solid waste from wet market of Bandar Baru Bangi Malaysia. *Australian Journal of Basic and Applied Sciences, 5*(5), 975–983.

17. Ojo, A. O., Akinbode, O. A., & Adediran, J. A., (2011). Comparative study of different organic manures and NPK fertilizer for improvement of soil chemical properties and dry matter yield of maize in two different soils. *Journal of Soil Science and Environmental Management, 2*(1), 9–13.

18. Saad, N. F. M., Ma'min, N. N., Zain, S. M., Basri, N. E. A., & Zaini, N. S. M., (2013). Composting of mixed yard and food wastes with effective microbes. *Journal of Technology, 65*(2).

19. Pathak, A. K., Singh, M. M., Kumara, V., Arya, S., & Trivedi, A. K., (2012). Assessment of Physico-chemical properties and microbial community during composting of municipal solid waste (*Viz. KItchen* waste) at Jhansi City, UP (India). *Recent Research in Science and Technology, 4*(4).

# CHAPTER 17

# Grub Composting: Experience on Black Soldier Fly Larvae (BSFL) Production from Organic Waste

S. ISMAIL,[1] M. S. M. ZAHARI,[1] A. N. MAZLAN,[1] A. ZULQARNAIN,[1] and N. A. UMOR[2]

[1]University Malaysia Terengganu, Kuala Nerus – 21030, Terengganu, Malaysia

[2]Faculty of Applied Sciences, Universiti Teknologi MARA Cawangan Negeri Sembilan, Malaysia, Phone: +601110910939, Fax: +603-8940-8319, E-mail: noorazrimi@gmail.com (N. A. Umor)

## ABSTRACT

Currently, there was an urge for a new way of waste management that was a more sustainable and smaller footprint. For this purpose cafeteria at UMT had been chosen as a model to study the method, namely, grub composting by using *Hermetia illucens* or black soldier fly larvae (BSFL) to convert food waste (FW) into compost and animal feed in a biowaste composter. This study involved an FW collection and grub composting experiment that last for two months. The composting was performed in biocomposter initiated by the ovipositing BSFL came from natural surroundings. It was found that approximately 700, 1000 and 1198 larvae harvested from different amount of waste for the first batch (800 g), second batch (1000 g) and third batch (2000 g), respectively. The findings revealed that the highest reduction rate of FW (g/day×larvae) was achieved by protein substrate with the addition of booster with the reading of 0.025 g/day×larvae. The addition of booster to the FW degraded by *H. illucens* encourages its reduction rate. Bio-composter proves to be a successful model in managing FW with the help of ecological engineered BSFL within the UMT area.

## 17.1  INTRODUCTION

Food waste (FW) continued to be among the highest waste generated in Malaysia, approximately about 31% to 45% of the total volume waste generated every day [14]. In addition, Solid Waste Corporation of Malaysia recorded that in 2015 alone, the FW reached up to 15000 tons everyday including 3000 tons that considered as untouched food [7]. Utilizing the FW in the composting process had becoming one of the sustainable methods for waste management [12].

Composting was the biological decomposition of organic waste such as food or plant material by bacteria, fungi, worms, and other organisms under controlled aerobic occurring in the presence of oxygen conditions [5]. In which in the end, the partially decayed organic matter called humus was collected. Interestingly another type of composting named as grub composting used black soldier fly larvae (BSFL) *Hermetia illucens L.* (Diptera: Stratiomyidae) to convert FW or manure, into compost for fertilizer and grubs for animal feed. The technique was sustainable as bioconversion of organic wastes by BSFL proven to be an effective method for degradation of FW compared to other method such as incineration, which will cost more as it requires higher energy and technology [16].

Black soldier fly (BSF) is a non-pest fly that can be found widespread through tropical and warmer temperate region between about 45°N and 40° and it is originated from southern USA [3]. Tomberlin and Sheppard [15] point out that BSF had a short life span and poses behavior that limit stay and encounter with human and their houses. Furthermore, BSF was a non-vector organism and reported to be able in animating *Escherichia coli* and *Salmonella* [4].

Life cycle of BSF [6] begin when the mating occurs between adult that took place in about two days after emergence and their ovipositing takes place after two days of fertilization [15]. Then female BSF would lay eggs near the food source to ease the larval to feed their self after hatching. It was the primary stage as the waste reduction occurred during this time. In normal condition, it took about 2–4 weeks for the larval development depending on the temperature and food availability. Figure 17.1 shows the life cycle of BSFL.

University and colleges are the forefront as model for new innovations, in particular for research in FW management that benefitted the communities around them all across the country [1]. In conjunction with this, a study was conducted to demonstrate grub-composting method to manage

FW of the cafeteria in University Malaysia Terengganu (UMT). Moreover, it was a stepping stone for the promotion of environmental sustainability in UMT. For this purpose, a small-scale prototype of bio-composter had been invented and constructed. Whereby this bio-composter was used to harvest the BSFL using grub composting. While the performance of FW reduction rate for grub composting using BSFL using different booster treatment were also analyzed.

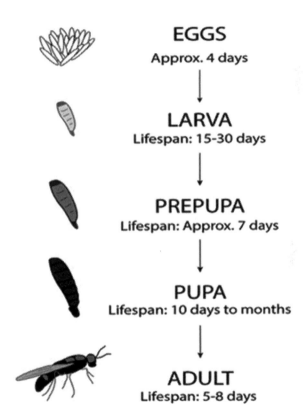

**FIGURE 17.1**   Life cycle of black soldier fly larvae (BSFL).

## 17.2   MATERIAL AND METHODS

The research was carried out to investigate the grub composting of FW from UMT`s cafeteria. All composting was done in bio-composter. The details of the procedures were presented in this section.

## 17.2.1   FOOD WASTE (FW) PREPARATION AND BSF OVIPOSITING

The waste collected from the UMT`s Cafeteria comprised of FW and kitchen waste. In order to attract the BSF, the FW was mixed with bio-enzyme and put in the Bio-composter that placed in a shady area. Few holes were fitted in the bio-composter so that the BSF would fly into the container and the females would be oviposited in the flutes of strips corrugated cardboard that was attached to the wall of the container [8]. After the eggs hatched, the larvae would fall onto their food source, which were the FW and started to feed. Larvae with more than 5-day old were collected from the container and washed with clean water.

## 17.2.2   COMPOSTING AND HARVESTING OF PREPUPAE BSFL

The composting was carried out in the bio-composter with configuration 0.4 m W x 0.6 mL x 0.4 m H (Figure 17.2). Three batches of composting were run with different amount of waste that was 800 g, 1000 g, and 2000 g, respectively. The composting continued for two months. Amount of prepupae harvested was determined in unit grams. After 60 days, the amount of prepupae harvested has been determined by using the sieving method in order to separate the residue/slurry. The following equation used for determination of larvae numbers and weight [11]:

$$\frac{\text{Weight of a larvae at 18 g}}{\text{Number of larvae at 119}} = \text{Weight of each larvae}$$

$$\frac{\text{Weight all larvae}}{\text{Weight of each larvae}} = \textit{Number of larvae}$$

## 17.2.3   REDUCTION RATE EXPERIMENT

The experimental method in this study was slightly modified from Song et al. [14] and Brams et al. [2]. For this purpose small container were prepared in which filled with three type of substrate namely 4 g of rice, 4 g of sausage and 4 g of fruits and vegetable in three replication. In each container, 30 larvae (hand-counted) that were more than 5-days old larvae were filled in. In addition, one container from each substrate was kept from larvae as control replicates. One of the containers filled with larvae with different

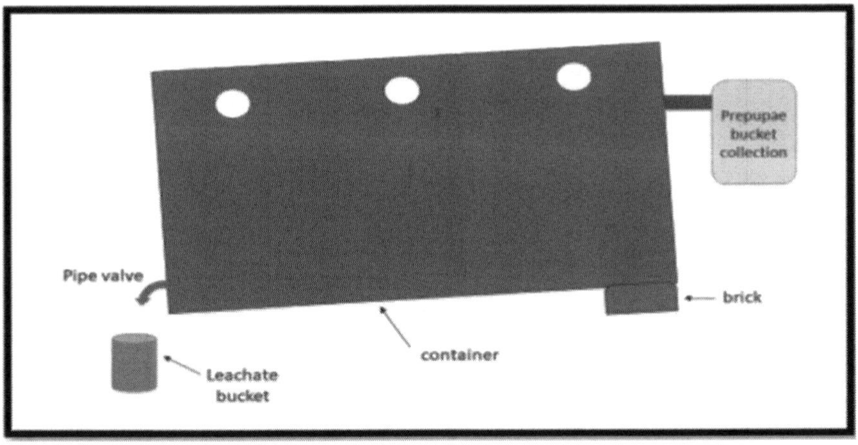

**FIGURE 17.2**   Bio-composter design.

types of waste was sprayed with foliar fertilizer that has been diluted with the ratio of 1 ml foliar fertilizer to 1000 ml of water. The containers were kept in dark box to protect the larvae from light disturbance. Then, to allow the good air circulating, the containers were covered with lids with nine holes on top of it. Mosquitoes mesh was clamped between the container and the lid to avoid oviposition by other flies in the container and to prevent the larvae from crawling outside the container [3]. The following equation was used to calculate the waste reduction rate [11].

$$WRR = \frac{(A - B)}{(D)(N)}$$

where; A = Dry weight of feed applied (kg); B = Dry weight of feed after (kg); C = Rearing Duration (days); N = Number of larva.

This equation was used to calculate the overall degradation (D).

$$D = \frac{W - R}{W}$$

where; W = Total amount of organic material applied during the time t; R= Residue after time t.

### 17.2.4   EVALUATION OF CONVERSION OF DIGESTED FEED

Conversion of digested feed (ECD) was evaluated to measure the effectiveness of BSFL in converting the digested feed into biomass [9]. The weight of the BSFL was measured before the feeding and after the feeding. The ECD was calculated using the following formula [11].

$$ECD = \frac{F - I}{A - B}$$

where; A = Dry weight of feed applied (g); B = Dry weight of feed after (g); F = Final body weight of BSFL (g); I = Initial bodyweight of BSFL (g).

## 17.3   RESULT AND DISCUSSION

### 17.3.1   PREPUPAE HARVEST

BSFL can digest and efficiently consume the FW in a bio composter designed. Figure 17.3 showed changes of FW during grub composting. There is approximately 700 larvae harvest for the first batch, with a FW weighing roughly 0.9 kg per batch. The biowaste composter progress for the first batch was disturbed with the presence of houseflies during first few weeks. However, with time passing dominancy of BSFL had excluded them. For the second batch, approximately 1000 larvae were harvested from FW weighing roughly 1 kg. While for the third batch, 1198 larvae were obtained out of 2 kg FW. The highest number of prepupae harvested was at third batch feeding that was influenced by the total amount of the feed. Table 17.1 presented data of prepupae harvest in three batch feeding.

Food waste before composting        During composting        After two month

Harvested BSFL

**FIGURE 17.3**   Changes of food waste in grub composting.

**TABLE 17.1** Prepupae Harvest Data

| Batch | Feed Source | Total Amount of Feed (g/day) | Residue + BSFL (g) | Number of Prepupae Harvest |
|-------|-------------|------------------------------|--------------------|----------------------------|
| 1. | Food waste (rice and fiber) | 800 | 315 | 700 |
| 2. | Mixed (rice, protein, and fiber) | 1000 | 220 | 1007 |
| 3. | Mixed food waste | 200 | 150 | 1198 |

## 17.3.2 WASTE REDUCTION INDEX

Figure 17.4 showed a graph of the waste reduction index by BSFL. According to Diener et al. [3], good reduction efficiency was indicated by high waste reduction index. WRI expresses the relationship of the time taken for the FW to degrade and the amount of the FW reduced. The WRI was found to be the highest when the BSFL was fed on the sausage with the addition of booster (18.28%). The lowest WRI value was found in the substrate of rice without being degraded by BSFL and no booster added (5.23%). The production of $NH_3$ and $CH_4$ due to the process of organic waste decomposition in aerobic condition could also inhibit the FW consumption process by the BSFL.

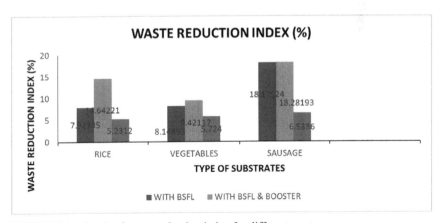

**FIGURE 17.4** Graph of waste reduction index for different waste.

## 17.3.3 FOOD WASTE (FW) OVERALL DEGRADATION

In this experiment, the overall degradation of the FW took place in 5 days. Table 17.2 showed sausage degraded by the BSFL with the addition of booster had the highest overall degradation with result of 0.9141 and the overall degradation was the lowest when there was no BSFL and booster added to the substrate which was rice (0.2616). The overall degradation was related to the type of substrate used, the used of the larvae to degrade the FW and the addition of the booster to the substrate. In most of the cases, substrate containing high protein would be preferable by BSFL [10].

Overall, the degradation of FW was higher when the FW was degraded by BSFL with the addition of a booster. As demonstrated by numerous studies, BSFL has a significant role in reducing the FW and organic waste volume and converting them into useful biomass [11, 14].

**TABLE 17.2**   Overall Degradation of Food Waste

| Type of Substrate | Overall Degradation (D) | Food Waste Reduction Rate (g/larva×day) | Waste Reduction Index (%) |
|---|---|---|---|
| Rice | 0.39619 | 0.01058 | 7.92384 |
| Rice + Booster | 0.73211 | 0.01953 | 14.64220 |
| Rice (control) | 0.26156 | – | 5.2312 |
| Vegetables | 0.40745 | 0.01106 | 8.14893 |
| Vegetables + Booster | 0.47106 | 0.01268 | 9.42117 |
| Vegetables (control) | 0.28620 | – | 5.72400 |
| Sausage | 0.90876 | 0.02445 | 18.17520 |
| Sausage + Booster | 0.91410 | 0.02456 | 18.28190 |
| Sausage (control) | 0.32668 | – | 6.53360 |

## 17.3.4   EFFICIENCY OF CONVERSION OF DIGESTED FEED (ECD)

The performance of BSFL in converting the FW into valuable biomass was evaluated by calculating the efficiency of conversion of digested-feed (ECD) as suggested by Muhammad et al. [9] and Stefan et al. [13]. According to Muhammad et al. [9], the ECD-feed indicates the effectiveness of BSFL in converting the digested feed into the biomass. Table 17.3 shows the weight changes of BSFL within five days of the experiment.

**TABLE 17.3**   ECD Value Data

| Type of Substrate | Weight of BSFL on 1st Day (g) | Weight of BSFL on 5th Day (g) | Weight Different of BSFL Within 5 Days (g) | Efficiency of Conversion of Digested Feed (ECD) (%) |
|---|---|---|---|---|
| Rice | 1.2350 | 1.3409 | 0.1059 | 6.67 |
| Rice + Booster | 1.2107 | 1.7963 | 0.5856 | 19.98 |
| Vegetables | 1.1225 | 1.2551 | 0.1326 | 7.99 |
| Vegetables + Booster | 1.1672 | 1.3748 | 0.2076 | 10.91 |
| Sausage | 1.259 | 2.6768 | 1.4178 | 38.66 |
| Sausage + Booster | 1.1859 | 1.6990 | 0.5131 | 13.93 |

BSFL fed with sausage produced the highest weight loss while the lowest was the BSFL that was fed with the rice. The ECD waste was found to be the highest in the sausage substrate (38.66%) and lowest in the rice substrate (6.67%). According to Stefan et al. [13], ECD value was in proportional with the food conversion efficiency. Although the FW reduction rate was also when the BSFL was fed on sausage + booster, the ECD value lower compare to BSFL fed with sausage only. This is to BSFL in the container did not consume enough substrate because of the anaerobic condition making its body gain less weight and affecting the ECD value.

## 17.4  CONCLUSION

At the end of the project, bio-composter was proven potential for harvesting the BSFL prepupae. Approximately 700, 1000, and 1198 larvae were harvested for the first, second, and third batch, respectively. Furthermore, these harvests were produced from different FW in every batch weighing at 0.9 kg, 1 kg, and 2 kg for batches 1, 2, and 3, respectively. It was also noted that the highest FW reduction rate was when the BSFL was fed on the sausage substrate with the addition of a booster with the reading of 0.02456 g/larva×day. This study however, concludes that the reduction rate of the FW is not proportional to the ECD feed. The highest ECD value in this study was 38.66% which was the sample of sausage degraded by BSFL. As a conclusion, BSFL shows a better potential as a prominent FW degradation method because of its ability to reduce organic waste in a short period and reduce environmental pollution. The addition of a booster to the grub composting process can also accelerate the FW reduction process.

## ACKNOWLEDGMENT

The author would like to express gratitude to the UMT administration for their excellent co-operation, inspirations, and supports during this project.

## KEYWORDS

- **black soldier fly larvae (BSFL)**
- **composting**
- **efficiency of conversion of digested**
- **food waste**
- **grub composting**
- **organic waste**

## REFERENCES

1. Babith, R., & Smith, S., (2010). *Cradle to Grave: A Study of Sustainable Food Practices in a University Setting.* Posters, Paper. http://opensiuc.lib.siu.edu/reach_posters/5 (accessed on 22 December 2020).
2. Brams, D., (2015). *Valorization of Organic Waste Effect of the Feeding Regime on Process Parameters in a Continuous Black Soldier Fly Larvae Composting System.* Swedish University of Agricultural Sciences Faculty of Natural Resources and Agricultural Sciences Department of Energy and Technology.
3. Diener, S., Zurbrugg, C., & Tockner, K., (2009). Conversion of organic material by black soldier fly larvae: Establishing optimal feeding rates. *Waste Management and Research,* (27), 603–610.
4. Erickson, C. M., Islam, M., Sheppard, C., Liao, J., & Doyle, P., (2004). Reduction of Escherichia coli O157:H7 and *Salmonella enterica* serovar enteritidis in chicken manure by larvae of the black soldier fly. *Journal Food Protection, 67,* 685–690.
5. Haug, R., (1993). *The Practical Handbook of Compost Engineering.* CRC Press. ISBN: 978087371373.
6. Jack, Y. K. C., Sam, L. H. C., & Irene, M. C. L., (2017). Effect of moisture content of food waste on residue separation, larval growth, and larval survival in black soldier fly bioconversion. *Waste Management, 67,* 315–323.

7. Lim, W. J., Chin, N. L., Yusof, A. Y., Yahya, A., & Tee, T. P., (2016). Food waste handling in Malaysia and comparison with other Asian countries. *International Food Research Journal*, (23) S1–S6.

8. Lydia, P., Shannon, J. C., Deb, N., Maurice, P., & Jean, S. V., (2018). Cultivation of black soldier fly larvae on almond by-products: Impacts of aeration and moisture on larvae growth and composition. *Journal of the Science of Food and Agriculture, 98*(15), 8–13.

9. Muhammad, Y. A., Robert, M., Adelia, F., Efrizaldi, A., & Isabela, R. H. S., (2016). Bioconversion of *Pandanus tectorius* using black soldier fly larvae for the production of edible oil and protein-rich biomass. *Journal of Entomology and Zoology Studies, 5*(1), 803–809.

10. Mutafela, R. S., (2015). *High Value Organic Waste Treatment via Black Soldier Fly Bioconversion*. Master thesis, Royal Institute of Technology, Stockholm, Sweden.

11. Nuraini, S. M. N., Jun, W. L., Man, K. L., Yoshimitsu, U., Thiam, L. C., Yeek, C. H., & Mardawani, M., (2018). Lipid and protein from black soldier fly larvae fed with self-fermented coconut waste medium. *Journal of Advanced Research in Fluid*.

12. Wahidah, A. G. S., (2017). *Food Waste in Malaysia: Trends, Current Practices, and Key Challenges*. Centre of Promotion Technology, MARDI, Persiaran MARDI- UPM.

13. Stefan, S., Naydenov, M., Vanchetva, V., & Aladjadjiyan, A., (2006). Composting of food and agricultural wastes. In: Oreopoulu, V., (ed.), *Book: Utilization of Byproducts and Treatment of Waste in Food Industry*. Publisher: Springer. doi: 10.1007/978-0-387-35766-9_15.

14. Song, Q., Lee, B., Tan, G. P., & Maniam, S., (2017). Capacity of black soldier fly and house fly larvae in treating the wasted rice in Malaysia. *Malaysian Journal of Sustainable Agriculture, 1*(1), 8–10.

15. Thomberline, J., & Sheppard, D. C., (2002). Factors influencing mating and oviposition of black soldier flies (Diptera: Stratiomyidae) in a Colony. *Journal of Entomological Science 37*(4), 345–335.

16. Van, H. A., (2013). *Potential of Insects as Food and Feed in Assuring Food Security Annual Review of Entomology, 58*, 563–583.

# CHAPTER 18

# Buffet Style and Food Waste Segregation Practice During Orientation Week: A Preliminary Study on Foundation Students of UMS

SITTY NUR SYAFA BAKRI, ELNETTHRA FOLLY ELDY, and
IZIANA HANI ISMAIL

*Preparatory Center for Science and Technology, University Malaysia
Sabah, Jalan UMS, Kota Kinabalu – 88400, Sabah, Malaysian,
E-mail: syafa@ums.edu.my (S. N. S. Bakri)*

## ABSTRACT

High volume of food waste has been consistently unavoidable in the event of large number of attendees. To avoid this and disseminate a proper food management, we have implemented a systematic food arrangement and food waste segregation practice during students' orientation week in Preparatory Centre for Science and Technology (PPST), UMS. We adapted a buffet style particularly for breakfast and lunch time, while conventional take-away method was maintained for dinner. During each breakfast and lunch session, an efficient food management was achieved by hiring different food vendors, which substantially reduced the meal time duration of 350 students into only 40 minutes. More importantly, single-use plastic waste was far less produced. Students were also been cooperative with the segregation of food waste. There is still room of improvement to reduce the waste production, particularly for single-use plastic waste. In the future, a systematic implementation of buffet style could be adapted for all meal session.

## 18.1   INTRODUCTION

Food is a product for human ingestion that provides nutritional support for a healthy growth [1]. When food is disposed or not consumed, it becomes a waste [2]. According to Ref. [3], generation of food waste (FW) represents poor efficiency of food management. Specifically, in large gathering event where great amount of food is served, generation of FW should be considered into food management. Other term to define food management is food literacy. It is a correlation between knowledge, skills, and behavior that able to plan, manage, select, and prepare food as well as eating food [1].

In University Malaysia Sabah (UMS), serving meal using disposable utensils such as paper or plastic cup, spoon, fork, and plate is a visible norm of food literacy in high number of audience programs. The reasons for using this option is because it is less time consuming, easy to manage and does not require a large workforce to wash the utensils during the ceremony. In particular, the programs involving enormous attendance is orientation week or *Minggu Suai Mesra* (MSM) where food is provided and often served with plastic wrap or plastic pack/container which is always accommodated by take-away food method. At this point, FW not only from the leftover meal, but also the single-use plastic from the packaging and serving. Note that during MSM of UMS, the audience number can increase to 1000 ~ 2000 people, while the normal dining time is within 2 hours.

Thus, in order to reduce the generation of FW in take-away method for high participants programs in UMS, we determine to practice buffet style for food serving during the MSM week by taking Preparatory Center for Science and Technology (PPST) as a preliminary study. In addition, we aim to apply FW separation as an additional practice in food management during MSM.

## 18.2   OVERVIEW OF THE PRACTICE

### 18.2.1   THE ORIENTATION WEEK OF PPST, UMS

PPST offers a foundation study in science with duration of one year for former high school students from science stream background. MSM or well known as MSM is an orientation day to welcome new students and marks the starting of an academic year of PPST. Most of the new students are 18 years old and up to 350 people per intake. During the MSM, foods are provided for every mealtime. Figure 18.1 illustrated the orientation week schedule that demonstrates the estimation of eating period for breakfast, lunch, and dinner. Prior to this pilot study, breakfast, and lunch will be served in buffet while dinner is take-away style.

| DAY / TIME | 8.30 AM – 4.30 PM | | | 2.30- 4.30 PM | | |
|---|---|---|---|---|---|---|
| DAY 1 | Registration | | | Briefing session with parents | | |
| DAY 2 | 8.30 - 10.00 AM | 10.30 AM - 12.30 PM | | 2.00-3.00 PM | 3.30-4.30 PM | 8.00-10.30 PM |
| | Academic talk | Success @ Campus | | | Safety @ Laboratory | Mentor-Mantee |
| DAY 3 | 8.30 – 10.00 AM | 10.10 -11.10 AM | 11.30-12.30 AM | 2.00-3.30 PM | 4.00 – 6.00 PM | 8.30-10.30 PM |
| | Student affair session | Library talk | Meet the counselor | Alumni gathering | PPST tour | Residential college events |
| DAY 4 | 8.30 – 10.00 AM | 11.00 AM -1.00 PM | | 1.00-1.30 PM | 2.00 – 4.00 PM | |
| | Rehearsal | Official closing ceremony | | Photo session | Beach cleaning | |

**FIGURE 18.1**    The orientation week schedule of PPST.

## 18.2.2   BEFORE MSM

In order to implement the waste separation during eating time, it is crucial for the food to be well arranged since the amount of the students is quite high. One of the ways is hiring different food vendor for breakfast, lunch, and dinner before MSM. In addition, arrangement for food collection is important so that the time for taking food is within the allocated period and goes smoothly.

### 18.2.2.1   HIRING DIFFERENT FOOD VENDOR FOR BREAKFAST, LUNCH, AND DINNER

All vendors were briefed beforehand where vendor that selected for breakfast and lunch will served the meal in buffet style while for dinner is take-away. For buffet, we asked the vendor to provide the spoon and plate as well as container for the used plate and cutlery after meal. We require the vendor to divide the rice, main course (chicken) and vegetables into two different boxes. For serving time, we asked the vendor to be on-site at least one hour earlier to serve the food either buffet or take-away.

### 18.2.2.2 ARRANGEMENT FOR FOOD COLLECTION FOR BUFFET AND TAKE-AWAY

Figure 18.2 show the flow of food collection for buffet that was proposed in this study. Two classrooms are set up accordingly and as illustrated in the diagram, the movement in and out during the food collection is one way. From the entrance, students will be forming two lines to avoid congestion while they took their food. The food will be served by MSM's students committee at ③ and ④ to ensure every student gets the appropriate amount of food without holding the line. The students continue to ⑤ and ⑥ for dessert and drinks. It is a free-seating concept including seating on the floor. When finish, students is expected to return the dishes and waste as label in FW segregation area from ⑦ to ⑫. For take-away, the arrangement is similar from ③ to ⑫ and set up for two.

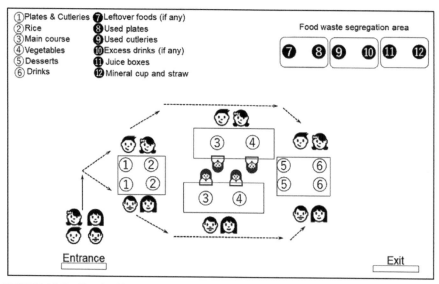

**FIGURE 18.2**    Food collection for buffet style.

### 18.2.3 DURING MSM

For buffet, the setup of Figure 18.2 was translated as in Figures 18.3 and 18.4 during the orientation week. Figure 18.5 shows the students formed two lines to collect food and assisted by MSM's student committee for main course and vegetable as in Figure 18.6. Figure 18.7 demonstrate the free sitting during meal time. On the other hand, Figure 18.8 show the FW collection area with labeled while Figures 18.9, 18.10, and 18.11 display bins for leftover meals, juices boxes and mineral cup and container for used plate and cutleries respectively. Figure 18.12 show the table and food setup for take-away.

**FIGURE 18.3**    Table setup for food collection.

**FIGURE 18.4**    Table setup for food collection with the food.

**FIGURE 18.5**    Students forming two lines to collect food.

**FIGURE 18.6**    MSM's student committee serving foods to the students.

**FIGURE 18.7** Students sitting accordingly to enjoy their meal.

**FIGURE 18.8** Food waste collection area.

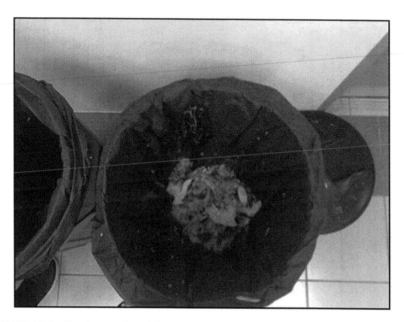

**FIGURE 18.9**    Food waste from leftover meal.

**FIGURE 18.10**    Food waste from juice boxes and mineral cups.

**FIGURE 18.11**   Container for used plates and cutleries.

**FIGURE 18.12**   Table and food setup for take-away during dinner.

## 18.3   THE IMPLICATION OF BUFFET STYLE AND FOOD WASTE (FW) SEGREGATION PRACTICE

From this study, we understand that despite the time constrain and large audiences, buffet style food serving can be implemented through a proper plan of food management. As a result, for 350 students, breakfast, and lunch took about 40 minutes to be completed through the buffet style of food serving during MSM of PPST. The time also including the FW segregation practice. Besides, similar period observed for take-away method during dinner. Here, we observed that when the vendor understands the flow of the food arrangement beforehand, they will easily give full cooperation to our requirement. In the same time, we also found out that by hiring different food dealer in every type of meal give no delay in delivery between the transitions of the mealtime.

On the other hand, FW showed that less leftover meals generated in buffet in comparison to take-away. This is because the students able to collect the amount of food that desired by their own. In addition, almost no excess of main course and vegetables were returned to vendor except rice. This finding expressed that generation of FW from leftover meals can be avoided when food is served in buffet style. In addition, less single-use plastic waste produced through buffet in contrast to take-away. Only single-use plastic from mineral cup and straw from juices drinks was observed and the tetra pak waste from the drinks packaging. For the take-out method, a high number of single-use plastic generated is coming from packaging of main course as well as the disposable cutleries.

### 18.3.1   SUGGESTION AND RECOMMENDATION

The outcome of this preliminary study allows us to propose several things for improvement:

1.   Firstly, since no issues on congestion during food collection, the food arrangement as illustrated in Figure 18.2 should be retained. In addition, dinner should also be served in buffet style for next MSM of PPST using similar arrangement.
2.   Secondly, to reduce the single-use plastic from drinks, instead of giving a tetra pak juices and mineral cup, juices, and drinking water, it can be served from water containers. For this matter, we suggest that, the students were given a reusable cup on the registration day

for their usage during mealtime. The cup then should be returned to PPST after MSM.

## 18.4   CONCLUSION

During MSM, generation of waste from eating is clearly understood by adding FW segregation practice in food management. The focus was not only on the FW but also on finding a good strategy in handling the flow of food collection for large crowd events. Well arrangement of utensil, food, and FW and hiring different food vendors for breakfast, lunch, and dinner were the great decisions made that lead to the successful buffet style in this study. In the future, dinner should also be served in buffet to reduce the generation of FW.

## KEYWORDS

- **buffet style**
- **food literacy**
- **food waste**
- **food waste segregation**
- **orientation week**
- **Preparatory Center for Science and Technology**

## REFERENCES

1. Vidgen, H. A., & Gallegos, D., (2014). Defining food literacy and its components. *Appetite, 76*, 50–59.
2. Göbel, C., Langen, N., Blumenthal, A., Teitscheid, P., & Ritter, G., (2015). Cutting food waste through cooperation along food supply chain. *Sustainability, 7*, 1429–1445.
3. Van, D. W. P., Seabrook, J. A., & Gilliland, J. A., (2019). Food for naught: Using the theory of planned behavior to better understand household food wasting behavior. *The Canadian Geographer, 63*(3), 478–493.

# CHAPTER 19

# Extreme Learning Machine in Laser-Assisted Machining Using Waste Palm Cooking Oil

F. YASMIN,[1] K. F. TAMRIN,[1] and N. A. SHEIKH[2]

[1]Department of Mechanical and Manufacturing Engineering, Faculty of Engineering, University Malaysia Sarawak (UNIMAS), Kota Samarahan – 94300, Sarawak, Malaysia, Tel.: +601115653090, Fax: +6082583410, E-mail: tkfikri@unimas.my (K. F. Tamrin)

[2]Department of Mechanical Engineering, Faculty of Engineering, International Islamic University, Islamabad, Pakistan

## ABSTRACT

The use of lubricating/cutting fluids is crucial in machining processes to reduce friction, alleviate heat accumulation, and prolong tool life. To minimize environmental and health impacts, a number of studies using vegetable oil-based cutting fluid have been investigated and reported demonstrating similar performance obtained using commercial cutting fluids. However, massive use of vegetable oil for such purposes would undeniably trigger issues of food security. In order to mitigate food waste (FW), the primary objective of the chapter is to demonstrate the application of waste palm cooking oil as a potential lubricating fluid in laser-assisted machining of metal. By considering kinematic and dynamic viscosities of the waste cooking oil, its effects on surface roughness and tool wear are studied by predicting using an extreme learning machine (ELM). The prediction results show that the average errors are only 0.51% and 1.19% for surface roughness and flank wear, respectively, suggesting good agreement between observation and prediction.

## 19.1   INTRODUCTION

The lubricants generally can be classified as automotive and industrial lubricants are used to improve the life of engines and machines by removing heat [1]. These lubricants, excessively used in industry and transportation, are largely produced as an expensive product of fossil fuel and also contribute significantly to pollution. The replacement of fossil fuels with bio-based oils and lubricants is not only environment-friendly, owing to biodegradability [2]. One major use of lubricants is in the automotive sector in different engine types at varying operating conditions. However, the performance of bio-oils as lubricants has not been much successful as yet owing to tribological issue resulting in deterioration of engine components [1].

While the other use of lubricants in industrial applications are mainly to protect the important operational elements from intensive duty cycles. The most important aspect of such lubricants is their viscosity to retain operations at optimal conditions irrespective of the ambient range of working conditions. Bio-oils have been used as machining lubricants, which helps in reducing cutting forces along with temperature reduction, which can improve tool life. However, one common issue with the mineral-based cutting fluids is the production of odor in the vicinity of the workplace, and a higher concentration of fumes can be hazardous for operators. On the other hand, the vegetable-oil based fluid is useful alternative since their biodegradable nature offers their renewability alongside excellent lubrication properties [2]. A few researchers have developed such cutting fluids. For instance, for the turning process, Katna et al. [3] used non-edible neem blended with various percentages (5%–20%) of food-grade emulsifier and showed promising results in terms of better-machined surface finish at 10% emulsifier blending compared to mineral oil. In addition, the rate of tool wear reduced notably while using neem blended with 5% emulsifier as cutting fluid in comparison with conventional mineral oil. While using palm oil as cutting fluid during high speed drilling of titanium alloys, Rahim and Sasahara [4] observed lower cutting forces and resultantly lesser temperature of workpieces compared to the use of synthetic ester as cutting fluid. Wickramasinghe et al. [5] used a novel coconut oil-based while performing end milling operation on samples made of AISI 304 steel. Results indicated that around 70% reductions in average surface roughness along with 48% reduction in flank wear in comparison with soluble oil as working fluid. In addition, Bermingham et al. [6] used laser-assisted milling of Ti-6Al-4V with vegetable-based cutting fluid to evaluate the life of the tool and the wear mechanism. The results suggested that the tool life improved around five times with respect to conventional dry laser-assisted milling. Such studies are underway with some hick-ups and limitations [7, 8]. Nevertheless, massive use of bio-oils extracted from edible crops such as palm and soybean has raised serious concerns on food insecurity.

On the other hand, wasting food is becoming a social problem along with its ecological and economic impacts. Globally around a quarter more than a billion tons of food is either lost or wasted [9]. Estimates of 2012 for Malaysia suggested that around 33 thousand tons of food is daily wasted, and this amount is on further rise owing to urbanization, changes in lifestyle, population growth, and industrialization. Large chunk of food wastes (FWs), especially of organic origin, is generated from common households, restaurants, and other industries in the food processing industries [10]. One peculiar item as waste, in millions of tons per day, is the cooking oil. Converting waste cooking oil of houses, restaurants, etc., to lubricants can help address one major environmental challenge since its disposal is a grave concern for countries across the globe [11]. To the best of authors' knowledge, no investigation has been performed on metal machining using waste cooking oil as industrial lubricant except automotive lubricants.

Different waste cooking oil would produce different machining characteristics and tribological properties. Characterization of waste cooking oil is necessary but laboriously time-consuming due to varying external factors. However, predicting the behavior of cooking oil in different engines and working environments is not an easy task. One important aspect is the use of prediction models, especially at the planning stage, to assess the behavior of tribological applications in finding suitable parametric ranges for better performance [12–14]. This approach can mitigate the assessment process complexities with lesser rejections and errors. With the benefits of artificial intelligence (AI), as noted in many fields including manufacturing processes, different variants such as artificial neural network (ANN), genetic algorithm, fuzzy logic, etc., can be efficiently used to model the process. One critical aspect of these techniques is the time consumption during the learning process in addition to generalization and overfitting [15]. For instance, ANN requires a certain set of data for generalization while overtraining can lead to overfitting [16]. One alternate is the extreme learning machine (ELM), which uses weights for the hidden nodes [17]. In addition, ELM is relatively simplified than ANN, while it reduces the time for train-test; moreover, it also provides better generalization in case of overfitting issues. In this regard, the comparison carried out by Mustafa [18] between the ANN model and ELM indicated that the ELM estimates are superior to its counterpart along with fast learning at a lesser number of iterations. Similarly, the work by Anicic et al. [19] demonstrated the same predictions. In general, ELM demonstrate high capabilities in all three areas of learning speeds, training errors, and norms of weights compared to other algorithms such as back-propagation [20].

Keeping in view the relevant literature in terms of exploring innovative methods and practices for mitigating food wastages, this chapter analyzed the effect of waste palm cooking oil on surface roughness of 6082 Al alloy and tool's flank wear width by predicting using ELM. Here, kinematic, and dynamic viscosities of the waste cooking oil were also estimated as input parameters for machining. The methodology is described in the following section.

## 19.2  METHODOLOGY

### 19.2.1  EXPERIMENTAL DETAILS

In this chapter, a medium-strength aluminum alloy (6082) having 6 mm thickness was used while undergoing laser-assisted high-speed milling. The selected material is corrosion-resistant with a melting point of 555°C, thermal conductivity of 180 W/m.K and hardness Brinell (HB) of 91. A detailed chemical composition of the selected material can be found in 2010 [21].

The setup for high speed milling operation included a cutting system along with lubricant fluid delivered in droplets and a heating system operating continuously by using laser power as shown in Figure 19.1. The heating system is positioned as top laser heating which is preferred during radial cutting owing to wider heating area; while the side heating position is efficient during axial cutting [22]. Here side laser-heating method is more preferable rather than top laser-heating method. The machine with numerical control for milling was X-Carve having a maximum speed of 8000 mm/min within the plane, while in the out of plane direction (z-axis), a maximum of 500 mm/min speed can be achieved. A 4-flute micro-grain carbide end-mill tool was used having 3.175 mm diameter and coated with aluminum titanium nitride (AlTiN).

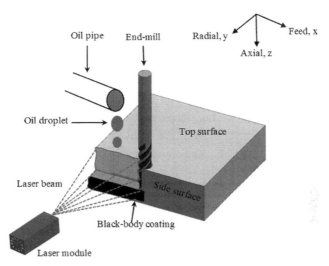

**FIGURE 19.1** Schematic diagram of the experimental setup.

The continual side-heating source was a diode laser having 1.6 W powers at a wavelength of 445 nm. The used beam spot of the laser had a spot size of 20 × 6 mm², approximately equivalent to the cutting area. To enhance absorptivity of radiations at the substrate alloy, the cutting zone was coated with blackbody coating. Before initialization of the cutting process, the substrate was pre-heated for at least 2 minutes.

In this chapter, waste or used sample of Malaysian palm cooking oil was explored as lubricating fluid during machining. The physiochemical properties are given in Table 19.1 for both off-the-shelf and waste palm cooking oil [23].

**TABLE 19.1** Physiochemical Properties of Bought and Waste Palm Cooking Oil

| Properties | Off-the-Shelf Palm Cooking Oil | Waste Palm Cooking Oil |
|---|---|---|
| Acid value (mg KOH/gm) | 0.3 | 4.03 |
| Calorific value (J/gm) | – | 39658 |
| Saponification value (mg KOH/gm) | 194 | 177.97 |
| Peroxide value (meq/kg) | <10 | 10 |
| Density (gm/cm³) | 0.898 | 0.9013 |
| Kinematic Viscosity (mm²/s) | 39.994 | 44.956 |
| Dynamic viscosity (mpa.s) | 35.920 | 40.519 |
| Flashpoint (°C) | 161–164 | 222–224 |
| Moisture content (wt.%) | 0.101 | 0.140 |

## 19.2.2   DESIGN OF EXPERIMENTS

The design of experiments was carried out for three important parameters involved in the machining process, namely, feed rate of cutting, lubricant flow rate, and the used laser power. In order to practice the process, several dry runs of the apparatus indicated the ranges of the selected parameters in terms of their results in producing the finished surface and the wear of the tool. At high-speed milling of Al 6082, feed rate range between 1500–3500 mm/min was selected with acceptable machined surface finish. In similar fashion ranges for laser power was selected between 650 mW–850 mW, while the ranges of cutting fluid kinematic viscosities are tabulated in Table 19.2 [23]. The quantity of cutting fluid was kept between 0.15 to 0.25 mL/s at pump voltage of 6.8 V to 8.3 V.

**TABLE 19.2**   Kinematic and Dynamic Viscosity of Palm Cooking Oil

| Input Parameter | | | Fluid Characteristics | | | |
| SL. No. | Power (mW) | Predicted Temperature Rise (°C) | Off-the-Shelf Palm Cooking Oil | | Waste Palm Cooking Oil | |
| | | | Kinematic Viscosity (mm²/s) | Dynamic Viscosity (mPa.s) | Kinematic Viscosity (mm²/s) | Dynamic Viscosity (mPa.s) |
|---|---|---|---|---|---|---|
| 1. | 0 at 40°C | 0 | 39.99 | 35.92 | 44.95 | 40.52 |
| 2. | 650 | 32.37 | 58.81 | 51.42 | 63.77 | 59.33 |
| 3. | 750 | 33.5 | 55.38 | 48.24 | 60.34 | 55.90 |
| 4. | 850 | 34.64 | 52.05 | 45.36 | 57.01 | 52.57 |

Using CCD (central composite design), the design of experiments was prepared with high and low levels of process parameters at setpoints of high (+1), central point (0) and low (–1). The resultant levels generate six axial points along with six central points with one replicate as summarized in Table 19.3, resulting in 20 experiments.

**TABLE 19.3**   Input Processing Parameters Based on CCD

| Parameter | Unit | Annotation | Extension | | |
| | | | –1 | 0 | +1 |
|---|---|---|---|---|---|
| Feed rate | mm/min | $F$ | 1500 | 2500 | 3500 |
| Laser Power | mW | $P$ | 650 | 750 | 850 |
| Droplet flow rate | mL/s | $r$ | 0.15 | 0.2 | 0.25 |
| Kinematic Viscosity | mm²/s | $v_k$ | 63.77 | 60.34 | 57.01 |
| Dynamic Viscosity | mPa.s | $v_d$ | 59.33 | 55.90 | 52.57 |

### 19.2.3  *EXTREME LEARNING MACHINE (ELM)*

In order to deduce conclusive evidence from the experimental setup and runs, a learning algorithm is employed. In this regard, ELM provides the linear and easy to implement learning machine as proposed by Huang et al. [17]. ELM improves the ability for generalization of feed-forward ANN by applying the learning structure to the hidden-layered.

In addition, the output weights are selected using Moore-Penrose generalized inverse in ELM while the initial guess of the weights as well as biases can be still arbitrary. The network training requirements are not necessarily present as linear equality, and is solved in a single step, which significantly reducing the time for relating $N$ number of input with output (input $x_N$ and output $y_N$) vectors.

The goal here is to relate the input process parameters with the output of the machining, as identified earlier, in the complete set of experiments. For instance, in an $i^{th}$ training of ELM, inputs $x_j$ are used to relate to responses $y_j$ with $j = 1, 2,..., M$. Here $M$ represents the total number of hidden layers defined as:

$$y_j = \sum_{j=1}^{M} \beta_j f\left(x_j, w_j, b_j\right)$$  (1)

where; $\beta_j$ are the weights of the output layer while $f$ represents the activation function. $w_j$ representing the weight vector connects the $j^{th}$ node of hidden and input with $b_j$ representing the bias of hidden nodes. Upon further simplifications, in the form of H, the final equation is expressed by Eqn. (3):

$$H = \begin{bmatrix} f\left(x_1, w_1, b_1\right) & \cdots & f\left(x_1, w_M, b_M\right) \\ \vdots & \cdots & \vdots \\ f\left(x_N, w_1, b_1\right) & \cdots & f\left(x_N, W_M, b_M\right) \end{bmatrix}$$  (2)

$$Y = H\beta$$  (3)

While the criteria for ELM is as follows:

$$L\left(X, Y; \beta\right) = Y - H\beta^2$$  (4)

Since ELM, unlike ANN, can seek output with the need for iterative procedure [24]; therefore the value for $\beta$ can be found using:

$$\beta = H^+ Y$$  (5)

The Moore-Penrose inverse of unique matrix $H$ is represented by $H^+$, also known as generalized inverse matrix.

In this chapter, using ELM, the relationship between inputs (process parameters) and outputs (surface roughness and flank wear) are built based on the above-mentioned data. The training data is categorized as 14 input sets/samples randomly selected from the total 20 data points while the rest six datasets were used to test data. Moreover, both inputs as well as outputs were initially normalized on a scale of −1.0 to 1.0 at the start of network training. Using MATLAB R2016a, ELM relied on a single user-defined value representing the number of hidden nodes, which was varied between 1 to 12, less than the total number of test samples [25]. While for the activation function, the sigmoidal function was used.

## 19.3   RESULTS AND DISCUSSION

After conducting the planned set of experimentations, each of the designated response variables were measured and examined which are listed in Table 19.4.

**TABLE 19.4**   Experimental Cases and Results Using Waste Palm Cooking Oil

| | Input Parameter | | | | | Machining Characteristics | |
|---|---|---|---|---|---|---|---|
| SL. No. | Feed Rate (mm/min) | Power (mW) | Flow Rate (mL/s) | Kinematic Viscosity (mm²/s) | Dynamic Viscosity (mPa.s) | Flank Wear (μm) | Surface Roughness (μm) |
| 1. | 1500 | 650 | 0.15 | 63.77 | 59.33 | 1.98 | 1.529 |
| 2. | 1500 | 850 | 0.25 | 57.01 | 52.57 | 1.70 | 1.499 |
| 3. | 3500 | 650 | 0.15 | 63.77 | 59.33 | 2.46 | 1.648 |
| 4. | 3500 | 850 | 0.25 | 57.01 | 52.57 | 2.02 | 1.612 |
| 5. | 1500 | 750 | 0.20 | 60.34 | 55.90 | 1.88 | 1.521 |
| 6. | 3500 | 750 | 0.20 | 60.34 | 55.90 | 2.25 | 1.627 |
| 7. | 1500 | 650 | 0.25 | 63.77 | 59.33 | 1.90 | 1.526 |
| 8. | 2500 | 650 | 0.20 | 63.77 | 59.33 | 2.04 | 1.546 |
| 9. | 3500 | 850 | 0.15 | 57.01 | 52.57 | 2.13 | 1.622 |
| 10. | 1500 | 850 | 0.15 | 57.01 | 52.57 | 1.77 | 1.513 |
| 11. | 2500 | 850 | 0.20 | 57.01 | 52.57 | 1.86 | 1.552 |

**TABLE 19.4** *(Continued)*

| SL. No. | Feed Rate (mm/ min) | Power (mW) | Flow Rate (mL/s) | Kinematic Viscosity (mm²/s) | Dynamic Viscosity (mPa.s) | Flank Wear (μm) | Surface Roughness (μm) |
|---|---|---|---|---|---|---|---|
| | | | | **Input Parameter** | | | **Machining Characteristics** |
| 12. | 3500 | 650 | 0.25 | 63.77 | 59.33 | 2.34 | 1.631 |
| 13. | 2500 | 750 | 0.15 | 60.34 | 55.90 | 2.02 | 1.542 |
| 14. | 2500 | 750 | 0.25 | 60.34 | 55.90 | 1.88 | 1.541 |
| 15. | 2500 | 750 | 0.20 | 60.34 | 55.90 | 1.97 | 1.534 |
| 16. | 2500 | 750 | 0.20 | 60.34 | 55.90 | 1.95 | 1.541 |
| 17. | 2500 | 750 | 0.20 | 60.34 | 55.90 | 1.94 | 1.538 |
| 18. | 2500 | 750 | 0.20 | 60.34 | 55.90 | 1.97 | 1.544 |
| 19. | 2500 | 750 | 0.20 | 60.34 | 55.90 | 1.96 | 1.550 |
| 20. | 2500 | 750 | 0.20 | 60.34 | 55.90 | 1.94 | 1.536 |

## 19.3.1 PREDICTION USING EXTREME LEARNING MACHINE (ELM)

Predictions using ELM for the surface roughness and flank wear using waste palm cooking oil, are shown in Tables 19.5 and 19.6, respectively. The hidden nodes were used for the prediction of surface roughness and flank wear having the smallest norm of the least-squares solution. In this case, the root mean square errors are minimum at 4 and 8 number hidden nodes for surface roughness and flank wear, respectively, where the average errors are only 0.51% and 1.19%, suggesting that the observation are in fine agreement with predictions.

**TABLE 19.5** ELM Prediction of Surface Roughness Using Waste Palm Cooking Oil

| Run | Feed Rate (mm/ min) | Power (mW) | Flow Rate (mL/s) | Kinematic Viscosity (mm²/s) | Dynamic Viscosity (mPa.s) | Experi- mentation (μm) | Prediction (μm) | Error % |
|---|---|---|---|---|---|---|---|---|
| | | | **Input Parameter** | | | **Machining Characteristics** | **Surface Roughness (Ra)** | |
| 1. | 2500 | 850 | 0.2 | 57.01 | 52.57 | 1.552 | 1.5321 | 1.28 |
| 2. | 3500 | 850 | 0.25 | 57.01 | 52.57 | 1.612 | 1.6191 | 0.44 |

**TABLE 19.5** *(Continued)*

| | Input Parameter | | | | | Machining Characteristics | | |
| --- | --- | --- | --- | --- | --- | --- | --- | --- |
| | | | | | | Surface Roughness (Ra) | | |
| Run | Feed Rate (mm/min) | Power (mW) | Flow Rate (mL/s) | Kinematic Viscosity (mm²/s) | Dynamic Viscosity (mPa.s) | Experi-mentation (μm) | Prediction (μm) | Error % |
| 3. | 1500 | 750 | 0.2 | 60.34 | 55.90 | 1.521 | 1.5166 | 0.29 |
| 4. | 3500 | 650 | 0.15 | 63.77 | 59.33 | 1.648 | 1.6343 | 0.83 |
| 5. | 1500 | 650 | 0.25 | 63.77 | 59.33 | 1.526 | 1.5248 | 0.08 |
| 6. | 2500 | 750 | 0.15 | 60.34 | 55.90 | 1.542 | 1.5402 | 0.12 |
| Average Error % | | 0.51 | | | | | | |

**TABLE 19.6**   ELM Prediction of Flank Wear Using Waste Palm Cooking Oil

| | Input Parameter | | | | | Machining Characteristics | | |
| --- | --- | --- | --- | --- | --- | --- | --- | --- |
| | | | | | | Flank Wear (VB) | | |
| Run | Feed Rate (mm/min) | Power (mW) | Flow Rate (mL/s) | Kinematic Viscosity (mm²/s) | Dynamic Viscosity (mPa.s) | Experi-mentation (μm) | Prediction (μm) | Error % |
| 1. | 2500 | 850 | 0.2 | 57.01 | 52.57 | 1.86 | 1.8515 | 0.46 |
| 2. | 3500 | 850 | 0.25 | 57.01 | 52.57 | 2.02 | 2.0306 | 0.52 |
| 3. | 1500 | 750 | 0.2 | 60.34 | 55.90 | 1.88 | 1.8117 | 3.63 |
| 4. | 3500 | 650 | 0.15 | 63.77 | 59.33 | 2.46 | 2.4387 | 0.87 |
| 5. | 1500 | 650 | 0.25 | 63.77 | 59.33 | 1.90 | 1.8875 | 0.66 |
| 6. | 2500 | 750 | 0.15 | 60.34 | 55.90 | 2.02 | 1.9996 | 1.01 |
| Average Error % | | 1.19 | | | | | | |

## 19.4   CONCLUSIONS

In this chapter, using ELM, the prediction of waste palm cooking oil as cutting fluid on laser-assisted milling operation of Aluminum alloy is carried out in terms of assessment of surface roughness and flank wear. Some important points can be drawn as a conclusion from this work as follows:

1.   The environmental concern on the impact of mineral oil/lubricants and their life cycle are significant, and many researchers are trying

to develop alternates using possibly renewable and lesser harmful vegetable oils for such purposes.

2. In the current chapter, the use of waste cooking oil in terms of its potential as an alternative cutting fluid/lubricant in metal machining process was presented. The results show promising aspect from the viewpoint of viscosity of such oil as lubricant.

3. ELM-based prediction errors of the surface roughness and flank wear were only 0.51% and 1.19%, respectively using waste palm cooking oil, suggesting good agreement between observations and predictions.

## ACKNOWLEDGMENTS

The authors are grateful to University Malaysia Sarawak for the Grant no. F02/TOC/1750/2018 allocated to the project.

## KEYWORDS

- **laser**
- **lubricant**
- **machining**
- **prediction**
- **tool wear**
- **waste**

## REFERENCES

1. Alotaibi, J., & Yousif, B., (2016). Biolubricants and the potential of waste cooking oil. In: *Ecotribology* (pp. 125–143). Springer.
2. Debnath, S., Reddy, M. M., & Yi, Q. S., (2014). Environmental friendly cutting fluids and cooling techniques in machining: A review. *J. Cleaner Prod., 83*, 33–47.
3. Katna, R., Singh, K., Agrawal, N., & Jain, S., (2017). Green manufacturing-performance of a biodegradable cutting fluid. Mater *Manuf. Process, 32*(13), 1522–1527.
4. Rahim, E., & Sasahara, H., (2011). A study of the effect of palm oil as MQL lubricant on high speed drilling of titanium alloys. *Tribology Int, 44*(3), 309–317.

5. Wickramasinghe, K., Perera, G., & Herath, H., (2017). Formulation and performance evaluation of a novel coconut oil-based metalworking fluid. *Mater Manuf. Process, 32*(9), 1026–1033.

6. Bermingham, M., Sim, W., Kent, D., Gardiner, S., & Dargusch, M., (2015). Tool life and wear mechanisms in laser-assisted milling Ti-6Al-4V. *Wear, 322,* 151–163.

7. Luo, Y., Yang, L., & Tian, M., (2013). Influence of bio-lubricants on the tribological properties of Ti6Al4V alloy. *Journal of Bionic Engineering, 10*(1), 84–89.

8. Kreivaitis, R., Gumbytė, M., Kazancev, K., Padgurskas, J., & Makarevičienė, V., (2013). A comparison of pure and natural antioxidant modified rapeseed oil storage properties. *Industrial Crops and Products, 43,* 511–516.

9. Martin-Rios, C., Demen-Meier, C., Gössling, S., & Cornuz, C., (2018). Food waste management innovations in the foodservice industry. *Waste Management, 79,* 196–206.

10. Hafid, H. S., Nor'Aini, A. R., Mokhtar, M. N., Talib, A. T., Baharuddin, A. S., & Kalsom, M. S. U., (2017). Overproduction of fermentable sugar for bioethanol production from carbohydrate-rich Malaysian food waste via sequential acid-enzymatic hydrolysis pretreatment. *Waste Management, 67,* 95–105.

11. Ng, H. S., Kee, P. E., Yim, H. S., Chen, P. T., Wei, Y. H., & Lan, J. C. W., (2020). Recent advances on the sustainable approaches for conversion and reutilization of food wastes to valuable bioproducts. *Bioresource Technology, 302,* 1–37.

12. Tamrin, K., Zakariyah, S., & Sheikh, N., (2015). Multi-criteria optimization in $CO_2$ laser ablation of multimode polymer waveguides. *Optics and Lasers in Engineering, 75,* 48–56.

13. Tamrin, K., & Zahrim, A., (2017). Determination of optimum polymeric coagulant in palm oil mill effluent coagulation using multiple-objective optimization on the basis of ratio analysis (MOORA). *Environmental Science and Pollution Research, 24*(19), 15863–15869.

14. Tamrin, K., Zakariyah, S., Hossain, K., & Sheikh, N., (2018). Experiment and prediction of ablation depth in excimer laser micromachining of optical polymer waveguides. *Advances in Materials Science and Engineering, 2018,* 1–9.

15. Ahmad, N., & Janahiraman, T. V., (2015). Modeling and prediction of surface roughness and power consumption using parallel extreme learning machine-based particle swarm optimization. In: *Proc ELM-2014* (Vol. 2, pp. 321–329) Springer.

16. Dashtbayazi, M., (2012). Artificial neural network-based multi-objective optimization of mechanical alloying process for synthesizing of metal matrix nanocomposite powder. *Mater. Manuf. Process, 27*(1), 33–42.

17. Huang, G. B., Zhu, Q. Y., & Siew, C. K., (2006). Extreme learning machine: Theory and applications. *Neurocomputing, 70*(1–3), 489–501.

18. Mustafa, A., (2018). Modeling of the hole quality characteristics by extreme learning machine in fiber laser drilling of Ti-6Al-4V. *J Manuf. Process, 36,* 138–148.

19. Anicic, O., Jović, S., Skrijelj, H., & Nedić, B., (2017). Prediction of laser cutting heat affected zone by extreme learning machine. *Optics Lasers Eng., 88,* 1–4.

20. Ćojbašić, Ž., Petković, D., Shamshirband, S., Tong, C. W., Ch, S., Janković, P., Dučić, N., & Baralić, J., (2016). Surface roughness prediction by extreme learning machine constructed with abrasive water jet. *Precis. Eng., 43,* 86–92.

21. Quintana, G., Gomez, X., Delgado, J., & Ciurana, J., (2010). Influence of cutting parameters on cycle time, surface roughness, dimensional error, and cutting forces in

milling operations on aluminum 6082 sculptured surface geometry. *Int. J. Machining Machinability Mater, 8*(3/4), 339–355.

22. Ding, H., Shen, N., & Shin, Y. C., (2012). Thermal and mechanical modeling analysis of laser-assisted micro-milling of difficult-to-machine alloys. *J. Mater. Process Tech., 212*(3), 601–613.

23. Ullah, Z., Bustam, M. A., & Man, Z., (2014). Characterization of waste palm cooking oil for biodiesel production. *International Journal of Chemical Engineering and Applications, 5*(2), 134–137.

24. Ucar, F., Alcin, O., Dandil, B., & Ata, F., (2018). Power quality event detection using a fast extreme learning machine. *Energies, 11*(1), 1–14.

25. Feng, G., Huang, G. B., Lin, Q., & Gay, R., (2009). Error minimized extreme learning machine with growth of hidden nodes and incremental learning. *IEEE Trans. Neural. Netw., 20*(8), 1352–1357.

# Index

Printed and bound by CPI Group (UK) Ltd, Croydon, CR0 4YY

23/10/2024

01777675-0010